A Modern Dredge in Operation.

DREDGING FOR GOLD
IN CALIFORNIA

BY

D'ARCY WEATHERBE

Member of the Canadian Society of
Civil Engineers

FIRST EDITION

San Francisco
MINING AND SCIENTIFIC PRESS
1907

COPYRIGHT 1907
BY
MINING AND SCIENTIFIC PRESS

AUTHOR'S ACKNOWLEDGMENT.

In completing the following work I wish to acknowledge the courtesy shown to me and the efforts that were made to facilitate my examinations among the dredging districts of the Sacramento valley, also for many valuable suggestions and data that have been furnished to me. Special acknowledgment is due to W. P. Hammon, George L. Holmes, W. S. and B. Noyes, Charles Helman, Newton Cleaveland, R. G. and Albert Hanford, John Plant, Karl Krug, O. W. Jasper, J. H. Leggett, H. Appel, F. J. Estep, and to many others.

SAN FRANCISCO, October 6, 1906.

D'ARCY WEATHERBE.

EDITOR'S NOTE.

Owing to his absence in South America, the author has been unable to read the proofs of this book. It has been my pleasant task to edit Mr. Weatherbe's manuscript, knowing how careful he has been in collecting the data on which this treatise is based; it is the result of judicious investigation by a trained engineer and it will commend itself to the profession.

T. A. RICKARD.

SAN FRANCISCO, April 22, 1907.

TABLE OF CONTENTS.

		Page
	Author's Acknowledgment	1
	Editor's Note	1
I.	Introductory	9
II.	Prospecting Dredging Ground	27
III.	Dredging Machines	46
IV.	Operation	88
V.	The Metallurgy of Dredging	109
VI.	Costs	139
VII.	The Horticultural Question	164
VIII.	General	170
IX.	Appendix	
	Contribution by J. H. Curle	183
	Gold Dredging, Editorial	188
	Sectional Dredging Machinery, Editorial	190
	Contribution by G. L. Holmes	192
	Contribution by C. W. Purington	194
	Contribution by D'Arcy Weatherbe	200
	Contribution by D'Arcy Weatherbe	206
	Cost of Dredging, Note	209
	Contribution by C. W. Purington	210
	Index	215

LIST OF ILLUSTRATIONS.

Fig.		Page
	A Modern Dredge in Operation . . . Frontispiece	
1.	Map Showing the Principal Dredging Districts . . .	11
2.	The Valley of the Feather River	12
3.	Hull of Dredge in Course of Construction	13
4.	Bed of Gold-bearing Gravel at Mississippi Bar . .	17
5.	Road-Cutting near Folsom	19
6.	Spill-Way of the Barrier Dam on the Upper Yuba . .	21
7.	The Barrier Dam on the Yuba	23
8.	Upper Feather River	24
9.	Marysville Rapid Transit	25
10.	Bringing a Keystone Driller into Place	29
11.	Prospecting Drill and Recovery Process on the Yuba .	31
12.	Rocker and Second Settling Vat	35
13.	Delivery-Vat, First Settling-Vat, and Sand-Pump . .	37
14.	Garden Ranch Dredge, Oroville	47
14½.	Diagram Illustrating the Development of the Dredge . .	48
15.	Garden Ranch Dredge. Dipper in Action	49
16.	Garden Ranch Dredge. Crane, Showing Gearing from Rear	50
17.	Exploration No. 3 Dredge Being Built on the Bank . .	51
18.	Frame of No. 3 Dredge, Folsom	53
19.	A-Shaped Forward Gauntree on the Ophir . . .	55
20.	Risdon Dredge at Fair Oaks	56
21.	Forward Gauntree on Boston No. 4	57
22.	Lattice-Truss Digging-Ladder of Yuba No. 8 . . .	58
23.	Digging-Ladder and Buckets on Oroville Dredge . .	59
24.	Rollers on Digging-Ladder, No. 3 Folsom	60
25.	Lower Tumbler of Folsom No. 4, Showing Wearing Plate	61
26.	Lower Tumbler Casting of Digging-Ladder	62
27.	Lower Tumbler on the Ophir	63
28.	Close-Connected Bucket-Line on the Butte	64
29.	Open-Connected Bucket-Line on the Baggette . . .	65
30.	Repairing Lower Tumbler of Exploration No. 2 . .	67
31.	Bucket of 13 Cubic Feet Capacity, on Folsom No. 4 .	68
32.	Bucket-Emptying Jets on the Butte	69
33.	Stacker-Discharge on Folsom No. 4	70
34.	Belt-Conveyor on Exploration No. 2	71
35.	Belt-Conveyor Stacker on Folsom No. 4	72
36.	Belt-Conveyor on Folsom No. 3	75
37.	Feather River Exploration No. 1, Showing Pan-Stacker .	77
38.	Motors and Ladder-Hoist Winch on Boston & California No. 3	79

LIST OF ILLUSTRATIONS.

Fig.		Page
39.	Sprocket-Gear Driving Arrangement on the Pennsylvania	80
40.	Pentagonal Equalizing Gear	81
41.	Folsom No. 3, Showing Standard Type of Drive-Gearing	82
42.	Wooden Spud of the Ophir	83
43.	Broken Steel Spud, Yuba No. 1	85
44.	Well of Yuba No. 8, Showing Tumbler, Grizzly, and Sluices	87
45.	Method of Anchoring and Moving the Boat	88
45½.	Cross-Section of Steel Spud	89
46.	Method of Anchoring a 'Dead-Man'	90
47.	Appearance of Bank when Spuds are Used	90
48.	Appearance of Bank when Head-Line is Used	90
49.	Chinaman Removing Trees in Front of a Dredge	91
50.	Removing Stumps Along Irrigation Ditch	92
51.	Movement of Boat by Side-Line	93
52.	Working Against Right-of-Way Stream	94
53.	Tail Sluices and Spuds on Yuba No. 7	95
54.	Folsom No. 5. Working High Bank with Water Jet	97
55.	Sand Pump Working on Yuba No. 7	98
56.	Plan Showing Cuts Across Pit, and Prospective Cuts	100
56½.	Daily Report of Dredging Department	101
57.	Bucket-Line Laid Out on Shore Ready for Yuba No. 8	103
58.	Accident to Bucket-Line on Yuba No. 2	104
59.	Cable Transport by Board Trestles	105
60.	Cable Transport by Barrel Pontoons	106
61.	Cable Transport by Forward Gauntree	107
62.	Tail Sluice, Showing Angle-Iron Riffles and Cocoa Matting	110
63.	Trommel, Stacker, Tables, and Sluice on Exploration No. 1	111
64.	Clean-up Apparatus and Riffles on the Butte	112
65.	Revolving Screen, and Tables on the Yuba No. 4	113
66.	Tables on the Pennsylvania	115
67.	Riffle Tables and Stream-Down Box on Bibbs No. 2	116
68.	Screen, Sluices, and Tables on Leggett No. 3	117
69.	Screen, Table, Launders, and Sluices (Holmes) on El Oro	119
70.	Screen and Tables on the Baggette	120
71.	Launder Delivery and Lower Tables on the Baggette	121
72.	Tail Sluice on the Baggette	121
73.	Sluice, Plate, Screen on Folsom No. 4	122
74.	Screens and Tables on Folsom No. 5	123
75.	Stacker and Remodeled Sluices on Folsom No. 4	125
76.	Trommel on Yuba No. 8	129
77.	Long Tom on Yuba No. 4	131
78.	Clean-Up Apparatus on Butte and El Oro	133
79.	Jets for Shaking Screen, on the Garden Ranch Dredge	135
80.	7½ Cubic Feet Bucket on Boston No. 4	139
81.	Same as Fig. 80, but Rear View	140

LIST OF ILLUSTRATIONS.

Fig.		Page
82.	Upper Tumbler of El Oro No. 1, Stripped of Wearing Plates	146
83.	Lower Tumbler, Showing Wearing of Cushion Plates	147
84.	Bucket Line of the Ashburton Dredge	148
85.	Idler Drum. Diameter 9 feet 6½ inches	149
86.	Showing Wear of Bucket Bottom and Pin	150
87.	Bucket Line, Showing Wear of Lips	151
88.	Wear Plates for Lower Tumbler	153
89.	Wear Plates with Lugs for Upper Tumbler	153
90.	Ladder Hoist on the Baggette	156
91.	Showing Wear on Lips of El Oro Buckets	157
92.	Belt-Conveyor on Stacker-Ladder of Folsom No. 3	163
93.	Eucalyptus Two Years After Being Planted on a Stack Pile	166
94.	Eucalyptus Just Planted on Stack Pile	167
95.	Flat Tailing Bed of an Old Dredge	169
96.	Remains of Old Workings on the American River	171
97.	Old Chinese Workings at Folsom	172
98.	Gravel Bank at Oroville	173
99.	An Old Miner and His Rocker	174
100.	Sketch Map of the Yuba Bottom	179
101.	The Treacherous Yuba Bottom	181
101½.	Middle Fork of American River	182
102.	Wreck of the Colorado-Pacific	190
103.	Feather River, with Dredges in Operation	213

I. INTRODUCTORY.

Gold dredging has won an important place in the industry of mining, and it has done so by right of hard-earned merit. Probably no branch of mining has come so prominently to the front in so short a time; by proving its commercial value it has opened to the financier an opportunity for safe and large investments.

Ten years ago there were no dredges working in California. Today there are fifty, either in operation or building in the districts of Oroville, Yuba, and Folsom, and it is probable that in no single instance is money being lost, while many of the dredges are clearing net profits of from $2000 to $5000 per month.

It is estimated that over $5,000,000 is invested in the business of dredging in the three districts of Oroville, Yuba, and Folsom, and the combined producing capacity in these three districts is estimated at an annual output of about 30,000,000 cubic yards.

Naturally in such a young and growing field there is tremendous diversity of opinion as to the relative advantage and value of the different methods in the practice and design of the machines and their parts. In the account presented herewith I have endeavored to trace the progress of the industry in California, and to give as many actual examples from practical experience as possible.

Geological and Historical.

The explorations of the United States Geological Survey have defined three great mineral belts in California. Three parallel lines are to be observed in the structure of the region; the first is coincident with the summit of the Sierras, the second is along their approximate base, say, from Visalia to Red Bluff and about fifty miles west of the first or main axial line, and the third, equidistant, drawn from the neighborhood of Clear lake to Kern lake, marks the eastern edge of the Coast Range, the shores of the Pacific forming its western edge at about the same distance. These lines divide the State geologically as well as physically. The Sierra Nevada is a belt of intrusive granite of ante-Cretaceous elevation, but of Triassic and Jurassic age, partially covered by important

Pliocene river deposits. These mountains are of great height and are the region of precious metal (and some iron and copper) mining, while the Coast Range, formed of strata chiefly of Cretaceous or Tertiary age and of post-Cretaceous elevation, yield principally quicksilver ores and carbonaceous minerals.

In the Sierra Nevada, volcanic activity is supposed to have ceased, while solfataric and seismic disturbances are still apparent in the Coast Range, the effect of the latter having been terribly illustrated in the recent earthquake and its disastrous consequences. The great valley lying between them is covered by recent alluvial deposits laid down on the bed of a fresh-water sea. It is to a part of the western border of this region that attention is to be drawn. The gravel deposits occur in every variety of texture from fine pipe-clay through sand and gravel to rolled pebbles and boulders, sometimes weighing tons. It is now generally accepted that they have been laid down by the action of a system of late Tertiary rivers which had nearly the same general course as the present streams on the west slope of the Sierra, but whose channels were wide and slopes greater.

The waters of these rivers, eroding the auriferous slates with their included quartz veins, concentrated the precious metals in deposits often 350 to 400 ft. wide at the bottom and sometimes several thousand feet wide at the top. Their depth now varies from a few yards up to six or seven hundred feet. Volcanic eruptions have, in places, covered these deposits with lava and tufa hundreds of feet thick. Denudation and erosion ensued, and the products of volcanic activity have sometimes been covered in turn with gold-bearing detritus.

Ages ago, long before the advent of man upon this planet, the western coast of the continent lay approximately along the line of the present Sierra Nevada range, then included in flat beds of shale and slate and on a level with the present great interior plains. The present buried river-beds had their origin in immense sluggish streams or sloughs of great length and extremely tortuous. These probably included watersheds as far north as the present Columbia river; they extended east into Utah and emptied into the ocean. Presently, however, a change occurred in affairs terrestrial and an uplift began to produce the Sierra Nevada mountains. These sloughs with their increased grades became roaring torrents, erod-

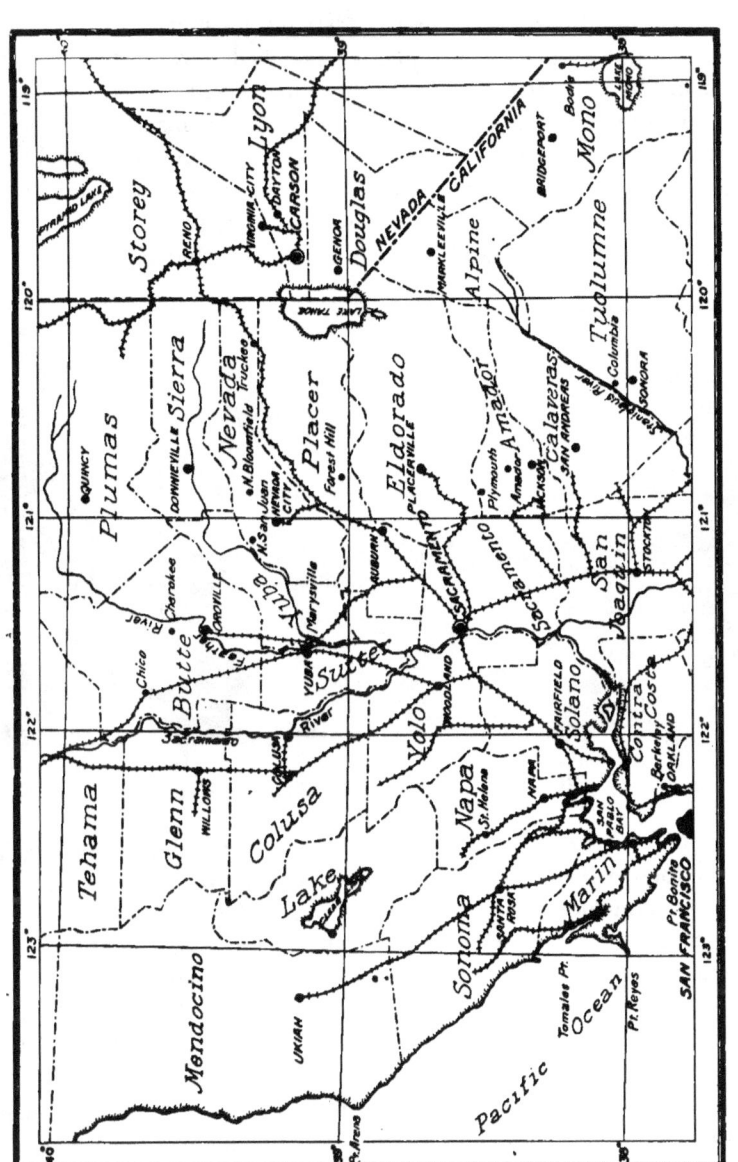

Fig. 1. Map Showing the Principal Dredging Districts in California.

ing deep channels through the embryo mountains and tending to straighten their sinuosities in the long and turbulent journey to the sea. Though enormous bodies of detritus must have been washed away, none of it remained in the river-beds owing to the fierce rush of the current. The upheaval which lifted the Sierras to an altitude greater than even that of the present magnificent peaks, abated and finally ceased; a period of erosion commenced. Up to this point it is probable that the river's burden was non-auriferous, but from the interior magma the slate and schist were penetrated by intrusive lava, followed by thermal activity that produced veins,

Fig. 2. The Valley of the Feather River.

the source of the gold in the placer mines of today. The Glacial epoch followed and during the rainy period that ensued, with the renewed influence of the sun's heat, tremendous erosion occurred; the peaks were worn down, and the ravines and valleys were formed. Simultaneously another phenomenon occurred that had an all-important bearing on mining. The valley of the Sacramento river, lying between the Sierra and the Coast Range, which had meanwhile become a vast inland fresh-water sea, began to rise. This uplift including the foothill country and its total vertical effect is variously estimated at from 800 to 1000 ft. The immediate result was a decrease in the grade of the great river system and the consequent slowing of the current, the channels becoming gradually filled with

Fig. 3. Hull of Dredge in Course of Construction.

gravel, sand, and silt, containing the concentrated gold from the eroded portions of the veins over which they coursed.

Contemporaneously, man appeared, and no written papyrus is necessary to record the event. Certain it is that he did not exist during the immediate post-Glacial period, but at this stage, when the rivers were filled with gravel, we have abundant evidence that he had even progressed in the arts. His prehistoric stone mortars for pulverizing maize or other cereals are constantly being brought to light in the hydraulic and drift mines, and piles of faggots and fire wood stacked by hand and exposed in the bench workings of the ancient rivers suggest the site of aboriginal habitations.

The final destruction, by burial, of the life of these channels came about through volcanic activity, which was at work only periodically during a comparatively short geological period; up to the time of the final outburst of molten lava, it appears to have produced mainly ash and dust that were swept down the streams in form of sand and slime. Subsequently this ejected material was compressed into the various layers of rhyolite and other tufaceous rock of different coarseness. Between these there were often beds of alluvium produced from detritus of non-volcanic origin.

There seems only to have been one general flow of actual melted lava, remnants of which may now be seen in the basaltic cap forming the isolated table mountains existing at several points in the foothills, notably on the Feather river and in Stanislaus county. This cap was in turn worn down by the later rivers, often flowing in an entirely different direction to the buried channels; these are thus found today at varying elevations up to several hundred feet above the existing rivers, which, cutting the ancient channels at many points, are locally indebted to them, as well as to the quartz lodes, for their gold. Indeed, it is probable that to this secondary concentration is due most of the enrichment at those favorable spots that provided the gold of the pioneers, and in fact all of the gold obtained in the pan, rocker, sluice, and other river workings. These streams of later origin continually changed their channels, though continuing to flow approximately in the same direction, and at each change they cut deeper. The result was that a series of terraces, bars and benches were formed, gradually rising from the present channel. A striking example of this action may be seen during almost any month in the year in the Yuba, but more emphatically

INTRODUCTORY. 15

shortly after the winter rains. Long sections of the river shift as much as several hundred feet in a few weeks, leaving gravel bars deposited in positions that a few days before represented the channel.

On the American river half a dozen distinct benches may be observed rising from the south side of the river for several hundred feet in height, and each of these contain consecutive auriferous channels—the former courses of the present river. It is these later river channels that in most cases form the richest portions of the dredging areas of today in almost every instance, though of course, quantities of gold-bearing matter must have been deposited by the ancient rivers previous to their becoming choked by the flow of volcanic material. That the gold from this source does actually exist at greater depths than it is possible to dredge (with present apparatus) has been proved by several comparatively deep bore-holes that penetrated below one or more beds of volcanic ash and also, I believe, by drift-mining operations in a few cases. Most of these drift mines worked by shaft in the valley are probably on the beds of the recent channels.

Several ingenious theories have been advanced to account for the source of the gold in the channels, but in no case has a reasonable argument been put forth. The 'marine idea' holds that these beds were deposited on the floor of a sea, but that this is altogether fallacious is shown by the fact that no marine remains have been detected and terrestrial signs such as trees and even human implements and (in some of the older beds) remnants of mammals, such as the mastodon, are found. The famous 'blue lead' theory pre-supposed the existence of an ancient river flowing from north to south and roughly parallel with the crest of the Sierras (some said this could be traced from Alaska to Mexico) and containing a characteristic blue gravel. In the first place, it has been proved that no such ancient channel exists and, secondly, that the 'blue lead' forms the bottom stratum of practically every buried channel, of Pliocene age, that has been opened. By analysis the 'blue lead' gravels contain iron pyrite, and these strata are similar in every way to the 'red leads' or 'rotten boulder' leads as they are called, which generally overlie them, except that in the latter case the iron has become decomposed and changed to an oxide.

That the gold is evidently derived from the bedrock traversed by the channel systems seems practically certain. The argument has frequently been brought forward that the veins throughout many of these districts are too poor to pay for working even under present economic conditions. This does not necessarily justify the conclusion that they were too poor to furnish gold to the channels. Some of the quartz veins have shown gold enough to warrant the investment of capital and they have even paid a profit. The $2 to $5 per ton that the average of these deposits contain is an extremely fine concentrate and represents the product of a small fraction of the mass of quartz broken and sluiced down these old rivers. Moreover, some of these veins have been eroded to depths approximating 1000 ft. and, incidentally, many fabulously rich pockets must have been treated by nature's concentrator—the river.

Placer deposits have been the earliest sources of gold throughout the world and since 1848, the date of the first important find in California, it is estimated that about four-fifths of the total output has been produced by the different forms of alluvial mining in this State. The total production from all sources up to date has been $1,450,000,000.

The successive steps in placer mining were the miner's pan, the cradle or rocker, the long tom, the riffle-box or sluice, the ground-sluice, booming or gouging, drift mining, hydraulic mining, the hydraulic elevator, and dredging.

The pan, rocker, and long tom are almost too well known to mention and may be passed over with but a word of comment. The pan was the earliest implement used in separating the precious metal from the accompanying gravel in California and is still necessary to the prospector, mill-man, and assayer. It is made of the best quality of Russia iron, generally stamped out of a single sheet, with the edge turned over a stout wire. The usual dimensions are: Diam. 10 in. at the bottom, 16 in. at top, and 2¼ in. deep. The angle of the sides is 37°. The method of use is as follows: The pan is filled with the gravel and sand and then carefully lowered under water; the fine and light material are gradually washed off, care being taken not to allow any gold particles to escape; the pebbles and coarser material, after examination, are removed by hand; washing continues until only magnetic sand

INTRODUCTORY. 17

and the gold remains. The pan being tilted and the water carefully manipulated, the gold forms a fringe at the top of the sand and is thus collected.

Fig. 4. Bed of Gold-bearing Gravel in the Old Workings at Mississippi Bar.

The 'rocker' is a box about 4 ft. long and 2 ft. wide, and is mounted on semicircular pieces of wood and worked by a handle to give it a side motion; and it is also inclined so as to carry the material down to the lower end, which is open. At the upper end is a small hopper that may be removed and which has a sheet-iron bottom perforated with ½-in. holes. Under the hopper is a canvas apron or tray inclined toward the head of the box but touching neither end of the hopper-box. A wooden riffle is placed across the box at the centre and another at the end. The material is fed into the hopper and screened through by water poured on top; the lighter material is carried over the end, while the riffles catch the gold and magnetic sand. This residue is panned at the end of the operation.

The 'tom' was originally a rough wooden box, 14 ft. long, and 2 ft. wide at the upper end, and 3 ft. at the lower end. The sides were about 10 in. high and the bottom had six or more cleats or riffles. The water was fed in a continual stream and the material treated was in larger quantities than the rocker. The next step was the 'sluice-box', which, as it could be lengthened indefinitely, had a much larger capacity. It is generally made in sections 12 ft. long and from 1 to 2 ft. wide. Riffles are placed both across and lengthwise with the box; mercury is introduced and the gold amalgamated.

'Ground-sluicing' came next; it consisted in bringing water through a ditch to a point above the claim high enough to produce a strong current. The bottom, if possible, is on bedrock and large stones form an artificial riffle. Occasionally a wooden sluice with riffles is placed at the end. Large quantities of material are shoveled from the sides into the ditch. The stones are finally removed and the concentrated material at the bottom is taken out and put through a rocker or long tom.

'Booming' or 'gouging' is somewhat similar to ground-sluicing, except that a large quantity of water is collected in a temporary dam above the workings and allowed to rush down suddenly, cutting away large quantities of 'pay dirt' which is conducted through ground-sluices.

The gravels were first tested by pan and rocker. The latter was introduced from Georgia, and toward the end of 1850 the long tom was brought into use. The dry bars of the rivers that were easy of access at low water, were first worked out; then the bottoms were worked, by using wing-dams; and finally entire

Fig. 5. Road Cutting near Folsom. Showing Hard Layers of Volcanic Ash and Gravel.

streams were deflected from their course by flumes and ditches. The lower 'bench gravel' was then worked by 'tom' and other sluices, booming, and gouging, etc. Gradually these benches and low alluvium supplies also becoming short, attention was turned in 1852 to the high hill deposits—the buried rivers. These immense alluvial banks, often capped with basalt, sometimes have a depth (or 'face') of over 600 ft., and as the richest stratum usually lies near the bottom, the grade of material was found to be, as a rule, lower in gold content than the deposits already described. A larger scale of operations and more economic methods became necessary. This was done by hydraulicking, the directing of jets of water under heavy heads against the bank, which is thus disintegrated and washed over sluices; it was also done by drift mining, following the deposits by 'tunnels' under the hills or reaching them by shafts on the flat lands.

Another method of working these deeper deposits, particularly where they do not contain water, is by the hydraulic elevator, an upright pipe into the bottom of which the gravel is drawn by a jet of water under head that also raises it to the top (sometimes 40 ft.), where it is sluiced in the usual manner. It is not, however, proposed in this article to enter into a discussion of the relative conditions surrounding the most advantageous use of each of the methods mentioned.

In the later 'seventies hydraulic mining had become one of the most important industries of California, in which more than $100,000,000 had been invested, in mines, reservoirs, canals, and equipment, but owing to damage done to farming lands in the valleys adjacent to the Sacramento river and its tributaries, this method of mining was inhibited in 1880 by Congressional act, and hydraulicking was almost entirely discontinued, until the passage of the so-called Caminetti Act, early in the 'eighties, which permitted the resumption of hydraulic mining under the permit and restrictions of a corps of United States engineers, known as the Debris Commission.

The Debris Commission's work is directed entirely to controlling the natural detrital flow, and although appointed for this work about 1883 and at present employing conscientious and efficient engineers, their whole work to date has been confined to the Yuba river, where practical results were only accomplished in

INTRODUCTORY. 21

1904. A point was selected a few miles above the present site of Hammon City, where it was proposed to place a dam, the duty of which was to arrest the burden of sediment in the river and collect it there. The height of the dam was to be sufficient for a year's accumulation and at the end of that period it was to be added to year by year.

It is commonly supposed that the efforts of the Debris Commission are to restore the practice of hydraulic mining throughout the State, but such is unfortunately not the case, nor does there

Fig. 6. Spillway of the Barrier Dam on the Upper Yuba.

appear to be the slightest hope for a largely increased output from this source in the near future. It is true that under certain conditions—the chief of which is that the miners provide impounding dams or settling basins for the tailing they produce—some approved mines may continue to work, but the formality to be gone through and the expense of this work as a rule prohibit large workings and the output from such mines as look after their own tailing is inconsiderable compared with the product of former hydraulic mining or what might be produced under the favorable mining conditions of today.

After several abortive attempts, the Yuba dam was built of piles and rubble, and capped with a sheet of concrete 18 in. thick. The horizontal distance between top of apron to end of toe of this first portion is 36 ft. and the height is 6 ft. It was estimated that the annual increase necessary would be about 8 ft.; but this has proved to be an over-estimate. In 1906 eight feet were added to the upper side, making the present total height from original base 14 ft. and the width from top of present apron to toe 56 ft. The length of the dam is 1200 ft. and a spill-way is now being constructed around the south end. Since the last addition the basin has filled up to within a few feet of the top with precipitated material, the water at the deepest point (near the spill-way) being about five feet.

Work of great magnitude too, in connection with this impounding problem is being carried on by the dredges and contract work at Hammon City, a few miles below the barrier dam. These settling basins, especially the one behind the barrier dam, should prove a valuable experimental ground for the State with regard to ascertaining the annual accumulation of gold brought down by the river. A series of borings, at close intervals across this and the other basins, through the material deposited there since their construction, would certainly be extremely interesting; at any rate it would give some data of the amount of concentration that had taken place in the early gravel flows.

An ambitious, though seemingly impracticable, suggestion has been made with regard to assisting the resumption of work by the hydraulic mines. It is proposed to direct the tailing toward the worthless tule lands, which cover a vast acreage in the Sacramento valley, and it has been contended that on filling these swamps with sand and gravel, they may become useful for agricultural purposes. This point has not been satisfactorily settled as yet, though at the present time it is understood experiments are being made with that end in view.

As stated above, the districts to be treated in this article are Oroville, Folsom, and the Yuba, as exemplifying the most modern dredging practice, and, as few mining camps are surrounded by such pleasant living conditions, a descriptive note may be appreciated in passing.

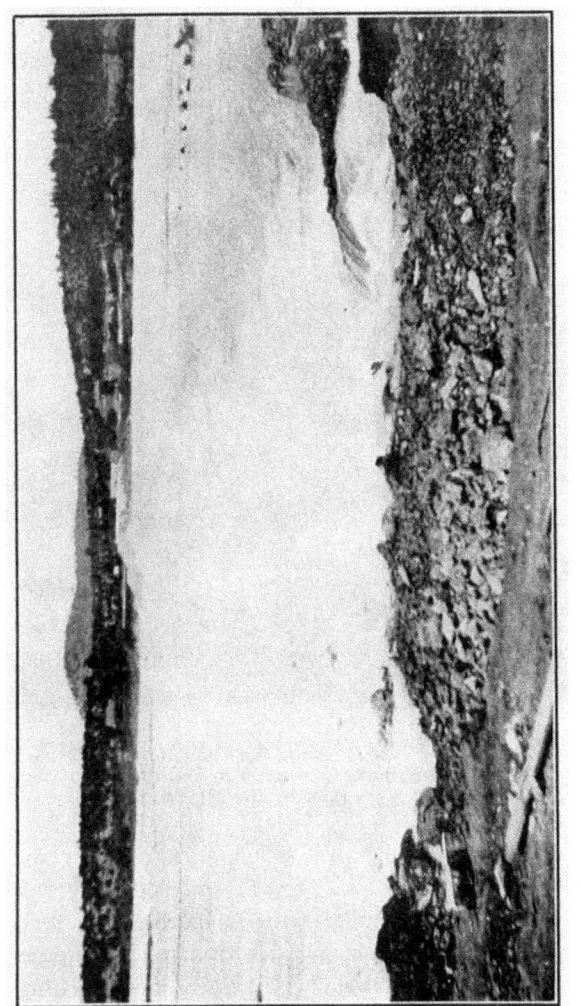

Fig. 7. The Barrier Dam on the Yuba.

To the stranger fresh from the East, the first view of the Sacramento valley is a new sensation. In his long trip across the continent he has no doubt seen prairie, desert, and magnificent mountain peak, but looking westward from the vantage point of the Sierras is a scene that for breadth and beauty cannot be surpassed. The prairie is usually rolling and its horizon is but a few miles away; the desert is deadly monotonous; here, however, is a refreshing change.

Fig. 8. Upper Feather River, showing the Dam Built by the Golden Feather Co. at an Immense Cost, only to find that the River Bottom had been Worked by the Miners of '49.

From the lava-capped heights rising abruptly near the headwaters of the Feather and Yuba rivers, one looks over a broad expanse of valley 50 to 60 miles wide. In the near foreground are miles of orchard land, bearing the orange, lemon, pommelo, peach, olive, and almond, interspersed with far-stretching vineyards or squares of wheat and grazing land. Silver streaks and patches shimmering in the bright sunlight denote rivers and overflowed levees. Halfway across the valley the jagged peaks of the Marysville buttes, a clear blue silhouette 20 miles long, thrust them-

INTRODUCTORY. 25

selves abruptly from the level plain to a height of 1800 ft. Further to the west, the Coast Range, a long low band of still fainter blue, capped occasionally by glint of sun on ice, may be just discerned, fading from sight as the eye traces it to the north. Where the Feather river escapes from the mountains and immediately below

Fig. 9. Marysville Rapid Transit.

the foothills, nestling among its orchards, lies Oroville. The name means "city of gold," and it is doubly applicable to the perennial color of its fruit and the precious metal contained in the soil on which it stands.

Marysville is the chief town and distributing point for the Feather and Yuba river dredging districts. It is one of the oldest towns in the valley and has the somnolent atmosphere of one of those communities in the Southern States not yet recovered from the effects of the war. Though this impression may belie the real commercial activity of its inhabitants, it is strengthened by the sight of the only public transportation facility afforded by the town—the mule cars—another ante-bellum reminder from the South. On the Yuba, 28 miles almost due south from Oroville, lies Hammon City, also situated in the foothills, and named from its founder, W. P. Hammon, the acknowledged leader of the dredging industry. The site of this settlement was only established about two years ago and though comprising but a handful of buildings, occupied by the dredging population, and temporary shops for repair work, it is growing fast and care is being taken to ensure its future beauty and convenience. The main street has been metalled with gravel and planted on either side with rows of black walnut and palm trees. Such older trees as were originally on the ranch have been left standing as far as is possible.

The town of Folsom, best known as the site of a State penitentiary and for its squalidness amid beautiful surroundings, is on the American river and marks the northeastern limit of the dredging ground on that stream.

II. PROSPECTING DREDGING GROUND.

In considering the prospective value of dredging ground there are many conditions to be taken into account besides its actual gold content; it is like all other classes of mines in this respect. As a prominent mining engineer recently said: "Each mine is a law unto itself." Likewise each dredging scheme must be considered strictly on its own merits before deciding on the methods and machinery to be used. Failure to recognize the specific conditions affecting a property, in calculating the means and cost of working it, has entailed the loss of large amounts of money. Misrepresentation and 'salting,' too, have played an important part in dredging as in other kinds of mining. Besides the amount of gold and the manner of its distribution, the following points must not be overlooked:

1. Characteristics of the gravel, as to clay, hardness, cementing, size of boulders, *et cetera*.

2. Depth to bedrock; the character and contour of the rock.

3. Permanent or variable water-level, and available water supply under head or otherwise.

4. Costs of power, labor, transportation, and supplies.

5. Another consideration, not seriously affecting dredging in the State of California, but to be taken into account in many other localities, is the climate.

The methods of determining the factors mentioned are by sinking shafts, by drilling, and by actual test with dredge. Many diverse opinions have been expressed as to the relative value of each method, but undoubtedly the most practicable is the shaft. Water, however, is a serious drawback to prospecting by shaft and therefore drilling is more common. The Keystone drill No. 3, boring a hole of 6 in. diam., is generally used in the Sacramento valley and it costs about $1700. It is a self-contained machine and consists of a walking-beam arrangement and an engine of 8 or 10 h. p. In drilling, about 52 strokes are made, while 54 strokes are made in driving. Casing is, of course, used and a shoe with steel cutting edge and weighing 800 lb. is placed on the bottom joint. The diameter of the shoe is 7½ in. outside and this diameter was used at first in calculating the area excavated.

Exhaustive tests have shown that the actual cubic content of the core brought up should be 0.27 of the linear depth bored, and though this is greatly modified one way or the other—depending on the nature of the ground—this result is theoretically true for the usual standard casing (which measures 5¾ in. inside diam.) and it has been arrived at by measuring the content removed from the sand pump after drying. Theoretically, provided that all the core drilled (and no more) is recovered, 16 in. should be drilled for each cubic foot of core produced. It will be found that this result varies from that obtained on the assumption that the diameter of the core is equal to the diameter of the outside of the shoe. The diameter should really be reckoned about midway between the outside and inside diameter of the drive shoe, thus allowing for the wear that takes place.

The practice varies as to keeping the casing below or above the bottom of the drilling. To prevent gold from a rich streak outside the area of casing being drawn in and thus making the computation valueless, the casing is kept 3 to 4 in. below drilling. On the other hand, if the casing is likely to strike a large boulder, or become plugged with clay, so as to prevent, for some distance, gold being extracted that really belongs to the core, then the drilling is kept a few inches below the casing. Probably the best method of obviating these difficulties is by drilling below the casing, but not pumping all of the material out until the casing has again been driven and drilling resumed. Conditions, such as hard or soft ground, old workings, etc., must govern this part of the practice so that sometimes it is found necessary to drive the casing much below the drilling.

The sand-pump is a hollow-steel cylinder 8 ft. long and 4 in. diam., with a valve at the bottom and a closely fitting plunger; on each side, near the top, it has an oval orifice to allow the pulp to be washed out. There are several methods of treatment, and the general practice is the same though the methods of keeping records and estimating results differ slightly. The apparatus used includes, a sluice-box 16 by 12 in. and 10 ft. long with holes at the lower end for allowing flow into a four compartment settling-tank, another settling-tank, and an ordinary rocker with several pans and tubs. A bucket is placed in the sluice and the contents of the pump are washed into this, each foot being

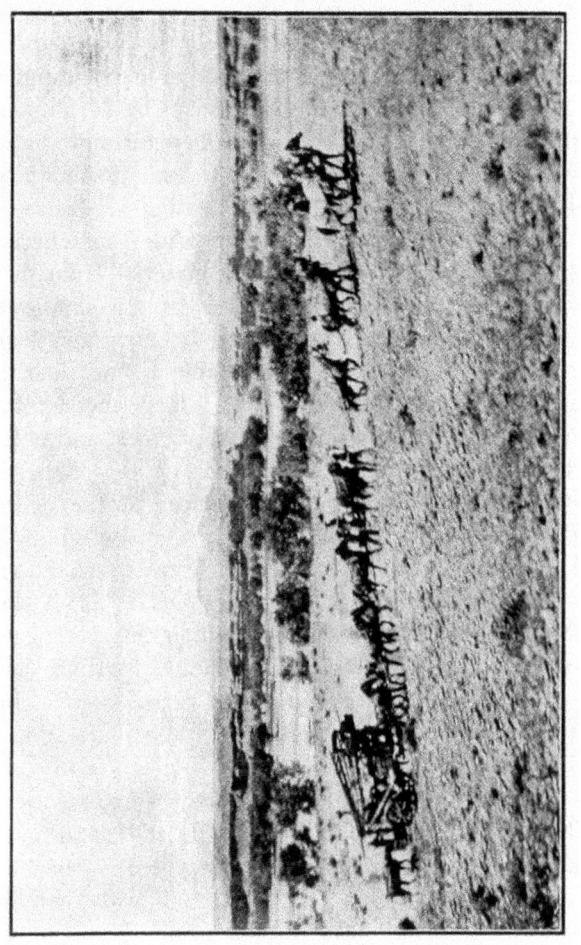

Fig. 10. Bringing a Keystone Driller into Place.

treated separately. The overflow runs off into the settling-tank. The content of the bucket is panned into a tub. The number and size of the 'colors' or particles of gold in each pan are estimated by eye and the result of each foot is noted in the log-book, the colors being classed in three sizes. The colors and black sand from each pan are kept separately; the former are segregated for each foot and then amalgamated. The gold is separated by nitric acid, washed, dried, annealed, and weighed, the resulting buttons from each hole being assayed for fineness. The surplus contents of the pannings caught in the tub are passed through the rocker, the concentrate being re-rocked. The overflow runs into another settling-tank and the contents of the two settling-tanks are roughly dried and measured, the computed yardage being thus checked.

Another method is as follows: The material from the hand-pump is received in a pan held in a sluice-trough similar to that mentioned above. The slime is allowed to flow away and the coarsest material is panned off in the tank. The finer portion of the material in the pan is panned over a large metal tub. The estimation of 'colors' is made in the usual manner, and everything in tub and trough is then passed over a rocker.

Often the rich material in the interstices of the bedrock or a rich seam will continue to be pumped into the hole, giving higher pannings than are warranted. To prevent this, when within a foot or so of the bottom, the results are panned as usual and from the foot just above and below bedrock the concentrate is caught in separate pans and if the results appear unduly high they are disregarded. All this is noted in the log-book, as well as the character and amount of gold content, and other particulars relating to the character of the gravel.

In one instance I observed an ingenious method of preventing salting. The work was being done on single shift, and on leaving for the night two panfuls of barren tailing were dropped down the hole. The heavy bit with rope attached was then lowered into the hole and the iron blocks were put on. To move these it would have been necessary to re-fire the boiler. In the morning the blocks were removed and the rope and bit washed into the hole and pounded to knock off any 'salt' (gold flakes) that might have adhered to them on being dropped down. The pans and sluice, etc., were thoroughly washed and the barren tailing at the bottom was tested. It might be a safer plan to keep a watchman on the spot.

Fig. 11. Prospecting Drill and Recovery Process on the Yuba.

A clever case of salting the drillings occurred recently at Oroville. The 'salter' mixed some finely divided gold with pipe-clay; he became so adroit that he could mark a piece of casing on the inside so that the streak would contain a given number of cents per cubic yard to be drilled. In case the driller noticed the marks at all, he would simply think that they were initials or shop numbers, and during the process of drilling, the streak with its fine gold was washed into the pump and unwittingly recovered with the gold in the gravel.

In connection with salting, several cases were related where the drillers wilfully produced (at the instance of their employer, it is said) misleading records by the following method: The bulk of the gold content was known to lie in a stratum 3 ft. thick and between 17 and 21 ft. from surface; each hole was drilled to about 30 ft., or 10 ft. deeper than was necessary. To illustrate the result we will assume that each of the 3 ft. between 17 and 21 ft. contained 60c. per cu. yd. and the cost of drilling was 7c. Thus the whole ground necessary to be dredged (20 ft. deep) would average 9c. per yd., leaving a profit of 3c. per yd. worked. The crafty party who attempted to wreck the negotiations for a fair sale, handed in his report, giving the depth of the ground as 30 ft., and the average yield per yard as 6c. Taking the vendor's admitted cost of working (7c.), there was an apparent loss of 1c. per yd.; and so the sale was declared off.

The following examples show the headings used in the log book kept by drillers of the Central Gold Dredging Co. and from the records of which the value of the property is computed. A summary statement is attached to the bottom of the report and is made up of details under the headings shown.

Date	Depth	Drilling Data					Gold Contents						Formation	Remarks
							Estimated Weight			Number of Colors				
		Drive	Core after Drive	Core after Pumping	Total Core for Drive	Drilled below Pipe	Coarse	Medium	Fine	Coarse	Medium	Fine		

A great deal has been said about the extreme care necessary in making estimates from drill-records. At the best, the method is but an approximation and the accuracy with which it is done should bear an exact ratio to the care taken in measuring the bank and calculating the cubic content dredged each month; otherwise, the results are not only misleading, but useless.

Thus the value per cubic yard of the ground at each particular hole is arrived at; the yardage of the ground necessary to be dredged is calculated from the depths to which the 'pay' is shown to exist, by the drill-records, and the actual average value per yard is determined by simply multiplying the value per cubic yard by the number of dredgeable cubic yards in the property. To estimate the purchasable value of the tract the first cost, depreciation, interest, and total ·cost per yard of operation must be deducted and, of course, a certain percentage allowed for the prospected value being lower than the recoverable value. One of the largest operators in the State told me that he always deducted at least 40% of the prospected value in purchasing dredging tracts.

As to the number of holes necessary to test the ground, this also is arbitrary. If the gold is known to be evenly distributed, the tract should be divided into five to ten acre squares. A flag at the centres of these marks the site of the hole. If the gold is in narrowing or widening channels, the ground should be first crossed by series of holes at long intervals and then closer together if favorable results are indicated.

The cost of drilling differs with the conditions, but it may be said that the price per foot will vary between $1 and $3.75 per ft.—a wide range. In one case at Oroville 13 holes were put down at a cost of $3.48 per ft., and in the same district at another property seven holes cost $2.40 per ft. Some contract work at Oroville was done for $2.50 per ft. The following recent examples of costs and speed cover a large area, and will give a practical idea of the variation: On the Yuba, five holes were drilled to an average depth of 93 ft. at an average cost of $3.85 per ft. The work was done during the winter rains and the roads were extremely heavy—freighting was difficult, transportation charges excessive, and there was great delay in having repairs made at Marysville (the nearest point), the fixed charges having to be sustained all the time. Moreover the work had to be finished within a time limit, so as to secure results before an option on the property expired.

In the southern part of the Oroville district 50 holes were put down averaging 35 ft. each and two drills were employed working simultaneously. The cost in this case averaged the extremely low price of 97c. per ft. The work was done in summer and all conditions, including soft ground, were most favorable. Water was close at hand and the fact that both drills ran under the same management reduced expenses. The labor included 1 foreman, 2 drillers, 2 helpers, 1 panner, 1 two-horse team and 1 teamster; the last also acted as water carrier.

At Oroville nine holes were put down about June, 1904, and totaled 258 ft. in depth. One machine was used on single shift and 33 days were employed in the work with 4 days more for moving.

The cost was divided as follows:

Labor	Per day.	Amount.
Driller	at $3.50	$129.50
Pumpmen	" 2.50	92.50
Firemen	" 2.50	92.50
Water Carrier and team	" 4.00	148.00
Panner		265.50
Hire of drill		133.33
		$861.33
Fuel		70.00
Supplies		68.60
		$999.93

Average number of feet per shift.................. 7.8
" depth of holes..........................28.7 ft.

Average cost per foot from time drill was hired
    to return (41 days).........................$3.88

In panning and for the boiler 800 gal. were used per shift. Cost of fuel per shift $1.90 (using coal).

In the following case a month's work is selected out of the seven months continuous work done just previous to June 1906 and in which 15 holes were bored to an average depth of 28 ft. During the month selected 5 holes aggregating 172 ft. of drilling were put down and cost $514.86 or an average of $3 per ft.

Fig. 12. Rocker and Second Settling Vat at Central Dredging Company's Drilling Plant.

Just 31 days were employed in the work and 5½ ft. were drilled per day. The above work was done in extremely wet weather and the ground was very tight. At some drilling operations in the Yuba bottom during the present summer, oil was used as fuel in the boiler and 55 gal. per day were consumed at a cost of about $1 per gal. The cost of prospecting a tract by drill-holes varies with the thoroughness with which it is done and for practical purposes the following simple calculation prepared in tabular form will show the range:

Cost of Drilling 100 Acres for Various Depths Up to 50 Feet.

Amount of Drilling 1 Hole per	For 20 ft. Cost per foot at				For 30 ft. Cost per foot at			
	$1.50	$2.00	$2.50	$3.00	$1.50	$2.00	$2.50	$3.00
	$	$	$	$	$	$	$	$
1 Acre	3,000	4,000	5,000	6,000	4,500	6,000	7,500	9,000
2 "	1,500	2,000	2,500	3,000	2,750	3,000	3,750	4,500
3 "	1,000	1,333	1,666	2,000	1,500	2,000	2,500	3,000
4 "	750	1,000	1,250	1,500	1,125	1,500	1,875	2,250
5 "	600	800	1,000	1,200	900	1,200	1,500	1,800

Amount of Drilling 1 Hole per	For 40 ft. Cost per foot at				For 50 ft. Cost per foot at			
	$1.50	$2.00	$2.50	$3.00	$1.50	$2.00	$2.50	$3.00
	$	$	$	$	$	$	$	$
1 Acre	6,000	8,000	10,000	12,000	7,500	10,000	12,500	15,000
2 "	3,000	4,000	5,000	6,000	3,750	5,000	6,250	7,500
3 "	2,000	2,666	3,333	4,000	2,500	3,333	4,766	5,000
4 "	1,500	2,000	2,500	3,000	1,875	2,500	3,125	3,750
5 "	1,200	1,600	2,000	2,400	1,500	2,000	2,500	3,000

With regard to the ratio of recovery between dredging and drilling much, of course, depends on the method and care employed, the relative efficiency of the gold-saving appliances and the manner in which the clean-up is accomplished. Most experienced operators contend that however carefully the operation is carried out (within the practical economic limits) and using the best of gold-saving appliances the results by drilling can never approximate the work of sampling and estimating an orebody underground. It might be supposed that by using an arbitrary method of (over or under) estimating the drilling results, a fair idea of the probable recoverable value might be arrived at and if the results bore a consistent relation (greater or less) to the recovery, this would be an effective method, but such is unfortunately not the case in practice. The

Fig. 13. Delivery Vat, First Settling Vat, and Sand Pump.

ratios are quite irregular and results vary both above and below the returns from prospecting. Only one general rule seems to hold, and its application is found to be generally true, namely, that when very high results—70c. to $3 say—are got by drill, the recovery from that place is sure to be lower. Likewise it is true in practice, that ground giving very low results from the drill, say from 1 to 5 or 6c., generally dredges considerably higher. Some experienced operators go so far as to say that drilling is practically useless except as a means of ascertaining whether gold is actually present in the ground or not. On the other hand at Folsom, where several thousand holes have been put down and the most careful records kept, it is asserted positively that an average approximating about 90% of the gold shown by drilling has consistently been obtained in dredging. Be that as it may, when it is explained that roughly a drill-sample will only represent something like $\frac{1}{200,000}$ to $\frac{1}{1,000,000}$ of the body of material to be worked, while in sampling a mine probably from $\frac{1}{5200}$ to $\frac{1}{10000}$ is taken of the orebody (which is not, as a rule, less homogeneous than the gravel beds), there should be at least a proportionate difference in the working results. Moreover, the method of estimation and selection cannot be compared for accuracy. The following authentic cases are extremely interesting in this connection: In front of the Boston & California dredge $No.$ 1, 22 holes were put down in one acre, and the results showed that the ground contained 60c. per cu. yd. When this particular acre was dredged, just 30c. per cu. yd. was recovered. In another instance, 25 holes were put down on one acre and the dredge recovered 95% of the estimated amount. On the Delancy tract, the results of dredging recovery came very close to the drill-estimate, the latter being very carefully done.

Several holes were checked by dredge at Oroville and the report illustrates the practical application of the rule laid down above, as follows:

Property.	Estimated value from drill-hole.	Recovered by dredge.
Gardella tract	$0.50	$1.25
" "	3.00	0.75
" "	0.00	0.30
Leggett tract	3.18	No record
Viloro	0.06	0.28
" "	1.44	0.30
" "	0.20	0.10
" "	0.12	0.30

PROSPECTING DREDGING GROUND.

Of course, the closer together the holes are put down and the more of them there are on a given tract, the closer the general result for the whole tract should come to the actual recovery; if the work is properly done, a fairly accurate idea of the value of the property may be obtained.

Month	Ground Dredged	Prospect Holes in and near acre dredged	Prospect Value	Recovery	Ratio of Prospect Value to Recovery
	Cu. Yd.	Number	Cents	Cents	Per Cent
January	20,340	2	15.35	15.38	100.20
February ..	32,000	2	15.35	10.65	69.57
March	41,160	2	37.70	18.78	50.00
April	50,760	3	37.70	18.91	50.00
May	47,700	3	16.25	16.94	104.24
June	40,750	2	10.35	9.09	87.96
July	44,380	2	11.45	9.62	84.02
August	39,300	2	11.45	10.87	94.93
September .	41,600	2	4.40	10.25	230.00
October	48,460	3	5.25	9.58	182.43
November..	40,300	3	17.23	8.37	48.58
December..	46,400	3	17.23	9.20	53.40

An average of about $2\frac{3}{8}$ holes per acre were drilled on the area dredged.

For the 12 months ending December, 1903, the *Biggs No. 1*[*] dredge of the Oroville Gold Dredging & Exploration Co., Ltd., worked 474,610 cu. yd., which, according to the estimate from prospecting, should have yielded 11.40c., the total recovery averaged only 8.45c., or 76% of the prospect value. The same dredge for the following 12 months of 1904 dredged 493,150 cu. yd., which yielded 12.32c. per cu. yd., whereas the prospect value of the area worked showed from the drilling tests an average content of 16.64c. per cu. yd., in other words, a recovery was obtained of about 74% of the estimated value by drilling. The following details of the above cases show that the ratios of recovery in cases of ground dredged after one or two drillings vary greatly, while the total average for the year appears to maintain a fairly uniform relation.

In another district in the Sacramento valley for the three months of March, April, and May, 1906, a certain dredge recovered an average of 25c. per cu. yd., where prospecting by drill had only indicated 18c. per cubic yard.

Naturally, careful shaft-sinking is a more satisfactory method of testing; it ascertains both the gold contents and the nature of

---

[*] The *Biggs No. 1* is now *Exploration No. 1*.

the ground far more efficiently and thoroughly, but the cost is sometimes prohibitive in wet ground. Otherwise it costs less than drilling. The so-called China shaft is the method usually employed and unless one has seen the work, one wonders how it was ever accomplished by hand. Probably workmen of no other nationality would do the work or, in fact, could work in such a narrow compass.

The shafts are sunk, circular in section, 3 ft. in diam., and the work is done by Chinese. Two men will do from 5 to 8 ft. per day; and at Oroville the contract cost is $1 per ft. Washing and estimating content will cost about 30c. per day more. These costs are for all work above water level. Below that level special arrangements have to be made and it is often altogether impracticable because 'John' is decidedly averse to working while water is being hoisted over his head, and little is he to be blamed for his objection. In more than one case he has sunk such shafts where it is dry, to depths of 40 ft. and over. In a large percentage of cases where water is supposed to be insurmountable, a centrifugal or some other form of pump could be installed with a small boiler and the work accomplished. A shaft 5 ft. square was sunk for a depth of 40 ft. within a few feet of the Feather river and on a level that at times was covered by the overflow. In this instance of successful prospecting, a 6 in. centrifugal pump (run by steam) was used. The cost should be little more for such work as electricity can now be successfully and economically used, and the results are so far ahead of the average drill-hole—if only for the purpose of positively knowing the nature of the ground—that were cost per foot several times more than drilling, it would still be advisable. The whole area of a certain property at Oroville could have been efficiently and cheaply tested in this manner, but for some inscrutable reason drilling was resorted to after a short period. Shaft-sinking on one property at Folsom cost $1 per ft. including panning, etc., and an average of 9 ft. per day was accomplished.

The only instances of work relating to the so-called paddock system that I am aware of in the Sacramento valley are those cases (cited in the *Bulletin* on 'Gold Dredging' published by the California State Mining Bureau) at Oroville, where Messrs. Hammon & Treat in 1895 sunk a pit about 100 ft. square down to the bedrock and used a centrifugal pump to keep the water out. The gravel was hauled in wagons to small sluice-boxes, where it was washed. In the other case the water was found to be too heavy to contend with on approaching bedrock. This work, however, was apparently done in a mining way and not as a prospect to test the ground.

Both at Oroville and Folsom most of the ground now being worked by dredge was all worked over years ago by pan, rocker, and sluice during the early days. As the grade of the gravel got lower, the white men left and Chinamen took their places and the ground was re-worked. In the dry season it is probable that these old shafts and drifts in many places reached the present bottom as shown by the timbers and portions of wing-dams that are encountered by the dredges and drills every day—much to the operator's disgust. Even in wet weather some of the ground was probably worked to the bottom with the assistance of the 'China' pump—a contrivance consisting of one or two 3 in. rubber belts with wooden blocks attached which acted as elevators. This was either driven by hand or by an overshot wheel of native manufacture. By this means in some cases the water was pumped from a depth of 40 feet.

Although not thorough, this work was so general and the district was so carefully exploited that today it is said that ground where old workings are not found is generally poor, and actual values of 15c. per pan and over have been obtained from some of the old tailing beds. These abandoned workings, when encountered by the prospect drill-holes, however, form a menace which often vitiates entirely the result of the hole, as it is impossible to say from what amount of gravel the results are produced. The following authentic record from some recent prospect holes at Oroville show how variable is the distribution of the gold. In some cases it seems to lie in well-defined streaks and elsewhere it is disseminated through almost every foot of the ground to be worked. The first and second records are of two holes on the El Oro tract at Oroville and the other two are from some drilling near the centre of the district, both on the same property.

On El Oro Tract.

Number and coarseness of colors.	Depth in feet.	Formation.
	1	Red clay.
	2	
	3	
	4	
	5	
	6	
1 fine	7	Large gravel with little red clay.
	8	
	9	
	10	
	11	Loose ground, probably old workings.
1	12	
3	13	Large wash.
15	14	
2	15	
1	16	
	17	
1	18	
	19	
2	20	
3	21	Large wash but very tight.
3	22	
1	23	
1	24	Solid but more clay and smaller gravel.
	25	
1	26	
1	27	
1	28	
1	29	
1	30	
1	31	
	32	
	33	
	34	
	35	
	36	
Trace	37	
	38	
	39	
	40	

Record of a Deep Hole on El Oro Tract.

Depth from Surface, Ft.	Colors	Remarks	Formation
9	0		Clay and soil.
11	11	Fine	Large wash and clay.
12	1	"	" " " "
15	3	"	" " " " and some sand.
29	{2, 3, 3, 5, 7}	"	Large wash and clay and some sand.
30	11	Large	Same as above but softer.
32	17	Medium	" " " " "
36	*Extra good		" " " " "
37	" "		Very large wash.
44	Good		" " "
45	0		Changing.
46	†5		Dead looking wash.
47	0		Sand with some clay.
66	0		" " " " & some boulders.
73	0		Clay, sand and mica.
78	0		Sand.
83	10	Very fine	Small gravel.
87	2	" "	Large "
91	0		Sand.
99	0		Small wash, tough clay.
100	0		Large wash.
107	0		Country rock.

*In this and the next two records below the number of colors was not kept but it was of better grade than the upper portion.

†Formation probably volcanic; colors supposed to have been carried down.

Case 1.

Depth		Estimated Weight in Milligrams			Number of Colors			Formation	Remarks
Ft.	In.	Coarse	Medium	Fine	Coarse	Medium	Fine		
3				1			3	Top-soil stiff and sticky.	
4									Very stiff
5									
6								Hard pan.	casing
7									
10									drove
12									
13								Little hard pan and fine gravel.	hard.
14									
15									
16								Some fine and medium gravel, much sticky clay and sand.	
17	3			0.1			1		
18									
19									
20	1			0.3			2	Much medium and coarse gravel.	Tight
21				0.1			2		
22				0.1			5	Little sticky clay and some sand.	and
23	2			1.4			4		
24				5.5			12		heavy.
26	½			9			15	Much coarse gravel and sand. Some clay and cement.	
26	11½			0.4			4		
27	11½			1			3		
29	2			1.5			4	Boulders.	
30									
31						?		Very coarse gravel much sand and a little sticky clay.	
32	1								Very
33	1½								
33	11							Coarse gravel, much sand and a little sandy clay.	tight.
34	9								

Case 2.

Depth		Estimated Weight in Milligrams.			Number of Colors			Formation	Remarks
Ft.	In.	Coarse	Medium	Fine	Coarse	Medium	Fine		
2	1			1			5	Topsoil and fine gravel	
3			1.3	1		2	2	Much medium and coarse gravel heavy. Some sand and clay.	
4							2		
5				.8					
6				1			3		
7				3			7		
8				.7			5		Soft
9				.2			2		
10				.2			1		tight
11	½		8	8.5		4	15		
12			2	3.5		1	7		and
13				2.3			8		
14			1.5	4		1	6		heavy.
15				.8			4		
16				.6			4		
17				.3			2		
18				4.6			8	Extra coarse here.	
19				.3			2		
20				1.3			3		
21				1			2		
22				2			5		
23				.5			4		
24				.5			2		
25				1.5			8	Extra coarse cemented gravel.	
26				.6			3		
26	8			.3			2		
28				1.7			10		
29				.1			1	Much sand	
30				.6			3	and	
31				.2			2	clay.	
32				.2			3		

III. DREDGING MACHINES.

Under this head it is intended to describe in a general way the construction of the several parts of a dredge, while the conditions under which they operate and the relative advantages of each are gone into more fully in the following chapters.

There is practically only one standard type of economically successful dredge in use today in the Sacramento valley and that is the endless bucket with stacker. Experiments in the shape of suction dredges (caissons and centrifugal pump) and dipper or steam-shovel dredges have been tried, but with the exception of the last mentioned (in some special cases), none of them have been successful. There are only two dipper dredges now working at Oroville and these will be described in detail.

Fig. 14½ shows the development of the California dredge, through the single and double-lift and New Zealand types, to the present machine.

A dredge is essentially a machine to excavate and recover the precious metal from gold-bearing gravel and as the transporting medium of the plant is water, a boat or scow is necessary for this purpose. Thus, a dredge consists of the hull, with its superstructure and housing; a digging ladder and chain of buckets; a disintegrating and screening apparatus; a system of gold-saving devices; pumps, anchoring arrangements, and a stacker for the disposal of the coarse portion of the material excavated; and the power-plant, consisting of motors, winches, gearing, etc.

The hull is built in scow form, the forward part being divided so as to form a well, in which the ladder and bucket-chain may be raised and lowered. As the main wearing parts—and, in fact, the bulk of the dredging machinery—are renewed constantly on account of wear and breakage, the hull should be built to outlast the original plant, and indeed to outlast the area which it is intended to work over. Great care should be exercised in the design to ensure the continued strength and stiffness so essential to the proper working of machinery. Timbers should be of large size and the main truss and frame-timbers should be full length without splice. In this country, wood (Oregon pine) is used entirely in construct-

Fig. 14. Garden Ranch Dredge, Oroville. Dipper or Steam Shovel Type.

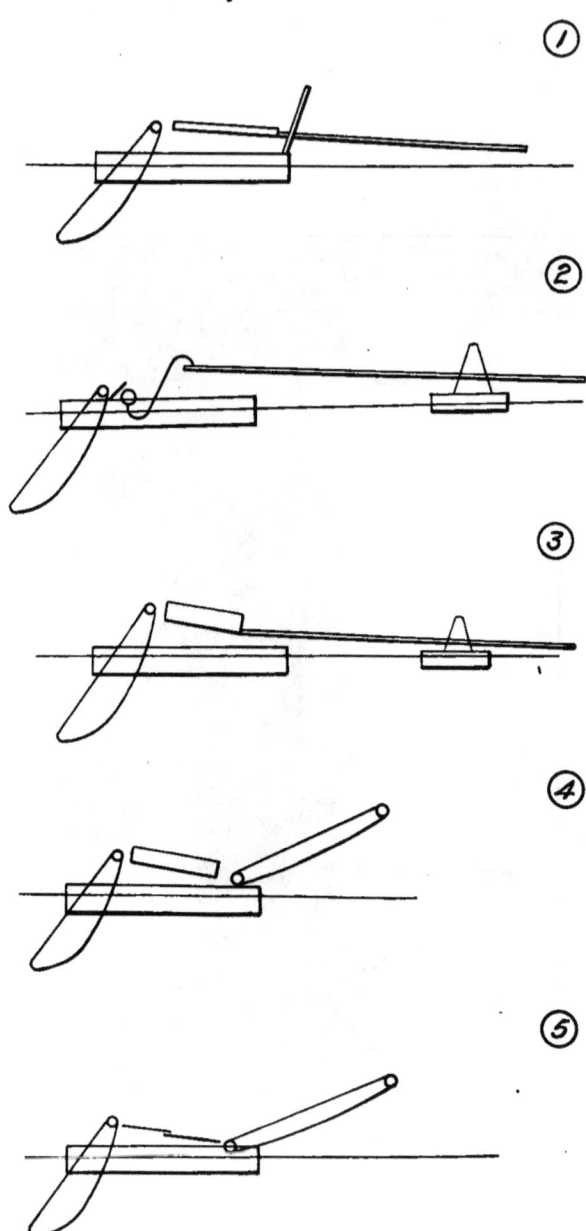

Fig. 14½. Diagram Illustrating the Development of the California Dredge.

DREDGING MACHINES. 49

ing the hulls, and steel for this purpose is not looked on with favor by dredging men.

As an example of the method and material of construction, specifications for one of the best of the Oroville boats are given as follows: The deck to be of plank 4 in. thick and 6 in.

Fig. 15. Garden Ranch Dredge. Dipper in Action.

wide; on the sides and bottom the planking to be 4 in. thick by 12 in. wide; the sides of the well and the bulkheads to be of plank 6 in. thick and 6 in. wide. The sides of the well are extended to the stern of the boat and form bulkheads aft and, in addition, there are two other bulkheads running the full length of the boat, all being securely fastened to the framing timbers of the hull, and firmly tied together by drift-bolts driven through the edge of the planking, making the whole bulkhead a unit in much the same manner as the web-plate of a girder. Lateral

Fig. 16. Garden Ranch Dredge. Crane, showing Gearing from Rear.

trusses in the hull and superstructure prevent the boat from sagging under the weight of the machinery. The head-beam is steel, built up of plates and angles, and the back guy connections are so

DREDGING MACHINES. 51

arranged that torsion is practically eliminated. All connections are carefully designed, and all bolts and rods are as large as experience has shown is desirable, and the washers are especially designed for this work.

Due regard must be given to the fact that in digging hard ground the tendency is to sink the stern and put the ends of sluices

Fig. 17. Exploration No. 3 Dredge being Built on the Bank.

under water, and the centre of balance is also changed when the ladder is raised or lowered. The hull is generally built on stocks or stilts in a pit previously dug, and this is filled with water after completion. Often, however, it is built alongside the pit and launched into it later. Each joint is filled with white lead, caulked, and the whole painted, inside and out; or it is treated inside with crude petroleum, which has been found an excellent preservative against water and dry rot. Hulls are always provided with ventilators, to prevent dry rot, and with sea-cocks, so that they may be sunk in case of fire.

After completing the hull and the framing of the gauntrees, etc., the water is allowed to flow into the pit and float the boat. It is then launched and is ready to receive the machinery, after the installation of which it is housed in. In floor-plan the design is rectangular, usually slightly rounded toward the bow and divided by the well-hole into two pontoons forward. The dimensions vary with the capacity of the boat and, to a certain extent, at the will of the designer. The following examples of large and small boats in the three districts will serve to give an idea of the difference in size.

District	Company.	Capacity of bucket. cu. ft.	Length. ft.	Hull. Width. ft.
Oroville	Feather River Exploration Co.	3¼	80	30
"	" " " "	5	80	32
"	Lava Beds Dredging Co.	5	84	30
"	Indiana Gold Dredging Co.	3	86	30
"	Boston & Oroville Mining Co.	4	88	30
"	" " " " "	5	88	30
"	Butte Gold Dredging Co.	3½	90	30
"	Indiana Gold Dredging Co.	3	92	34
"	Oroville Dredging Ltd.	5	94	36
"	Lava Beds Dredging Co.	5	96	36
Yuba	Yuba Consolidated Gold Fields.	7½	115	40
Oroville	Oroville Dredging Ltd.	5	110	40
Yuba	Yuba Consolidated Gold Fields.	7½	115	40
Folsom	El Dorado	7	96	42
"	Ashburton Mining Co.	7½	110	50
"	Folsom Development Co.	8.9	120	49
"	" " "	13	102	58

The depths of the hulls vary between 6 and 7 ft. and the draft from 3 to 5½ ft. A slight camber provides drainage. Most hulls have an overhanging deck of about 3 to 6 ft., which allows room for blacksmith's shop, etc. In one case a small air-compressor for riveting, etc., is installed.

Three frames, called 'gauntrees', are erected on the hull to support different parts of the machinery. The middle gauntree

Fig. 18. Frame of No. 3 Dredge, Folsom.

supports the upper tumbler and driving-gear and the upper end of the digging-ladder, though the bearing of the last is separate from the tumbler-bearing except in some of the later boats. The chain of digging-buckets is suspended from the upper tumbler and passes round the lower tumbler at the bottom end of the ladder.

The bow or forward gauntree is used for hoisting and lowering the digging-ladder and, incidentally, the landing-stage. The stern or after gauntree supports the stacker and provides guides for the spuds, if they are used. The gauntrees are, as a rule, of wood, though the Risdon company construct their middle gauntree entirely of steel. A number of the late designs of bow gauntrees have steel cap-pieces and, in most cases, steel guy-rods are used for bracing the frames. It is generally recognized that the type of bow gauntree with four parallel uprights is the best, as in those frames designed in the shape of the letter A the pontoons spread and tend to sink on the inner side. Much, too, depends on a proper form of bracing between the bow gauntree and the superstructure behind. This and the general methods of framing will be appreciated on consulting the accompanying photographs: Fig. 19 shows the A-shaped gauntree. Fig. 20 shows the typical Risdon design with four parallel uprights. Fig. 21 shows a steel-cap frame and a method of solid bracing behind.

The digging apparatus includes the ladder, the bucket-line, with the buckets and their parts, that is, lips, hoods, bottoms, pins, bushings (and links, in the case of an open-connected line), rollers, and tumblers.

The ladders and bucket-lines vary in length, being regulated by the depth of ground in which they are working. The former is built of steel, though in one of the original boats (now defunct) the ladders were of wood. At Oroville, where the average depth of gravel is about 30 ft. below water-line, they are from 60 to 90 ft. in length, and on the Yuba the longest ladder is 114 ft. between centres of upper and lower tumbler. The construction also varies according to the fancy of the designer; and they are generally formed of girder-sides built up of plates and angles, and braced across the top and bottom. A common form is that shown in Fig. 15 where the sides are built of angle-irons and steel plates with alternate plates and spaces across the top and bottom; the top and bottom plates being connected by another plate at the upper end to form bulkheads to prevent accumulation of dripping mud and gravel. Latticed braces of angle-iron are also used. Truss-rods are often fastened from each end on

Fig. 19. A-Shaped Forward Gauntree on the Ophir.

the bottom to give extra strength. Another form of ladder (Fig. 22) is braced across the top and sides with lattice-work of angle-irons instead of solid plates, and still another form built of angles with plates on the sides, has a continuous plate from top to bottom on the upper side (See also Fig. 23) to allow the surplus dripping dirt and water to sluice itself to the lower end, where it may be

Fig. 20. Risdon Dredge at Fair Oaks, Bucket 7 Cubic Feet, Electrically Operated.

again picked up by the buckets. It is doubtful whether this arrangement proves practical, particularly when digging deeply, as once the dirt strikes the water, the flow—particularly of the fine material—is impeded. At the top and bottom ends of the ladder steel castings are built in, to support the frame and carry the lower tumbler.

Rollers are provided on the upper side at intervals of from 5 to 7 ft. to reduce the friction of the bucket-line and prevent excessive wear on the ladder. Fig. 24 shows a digging ladder fitted with rollers and ready for erection. These are generally of cast iron, sometimes hollow and sometimes solid. In a few cases manganese steel has been used.

Fig. 21. Forward Gauntree on Boston No. 4 showing Steel Cap and Method of Bracing in Rea,r.

The tumblers are the heavy revolving castings at the top and bottom ends of the bucket-line round which the chain of buckets revolves. In section they are either square, pentagonal, or hexagonal, and the shape is governed largely by the theory of the designer. In Risdon dredges the upper tumbler is always square and the lower tumbler is generally pentagonal, the idea in having them different being to avoid the resultant jerk caused by the buckets leaving the peaks of upper and lower tumbler simul-

Fig. 22. Lattice-Truss Digging Ladder of Yuba No. 8.

Fig. 23. Lattice-Truss Digging Ladder and 7½ Cubic Feet Buckets on Latest Type of Oroville Dredge.

taneously; as a matter of fact, the jerk is largely caused by the fact of having a square upper tumbler instead of one of five or six sides. With the deep and narrow Risdon bucket, however, it is more or less necessary to use an upper tumbler of square section to insure complete dumping of the contents into the hopper. The irregular wear on pins and bushings soon obviates the jerk caused by tumblers of the same section by elongating the

Fig. 24. Rollers on Digging-Ladder, No. 3 Folsom.

bucket-chain; this wear affects the length of chain to such an extent, that usually from one to three buckets have to be removed after a short time to take up the slack.

The question of increasing the number of sides in tumblers so that they more nearly approach a circle in section has been discussed many times, and the question has now been pretty well settled by practical experiment. The number of sides must remain limited for two reasons: 1. In the case of the upper tumbler, after increasing the number of sides to six, its essential duty of holding, pulling round, and dumping the bucket-line is impaired and no practical solution in the shape of a sprocket

Fig. 25. Lower Tumbler of Folsom No. 4, showing Wearing Plate.

arrangement, any more than is now formed by the lugs and bottoms, as has been suggested, has been evolved; nor is it likely to be, on account of the immensely increasing weights and consequent strains set up. 2. The objection to the lower tumbler-section being increased to more than six sides or seven as a limit, is chiefly on account of the slippage and consequent wear.

The chief considerations that control the shape of the upper tumbler, however, are those which make for the most efficient and thorough emptying of buckets when dumping into the screen-hopper, and the least wear on pins and bushings. To aid this, jets of water are often used, playing into the full bucket as it rotates over the upper tumbler.

Fig. 26. Lower Tumbler Casting of Digging-Ladder.

The tumbler is not only constructed of different sectional shape but of different design. Some of the old patterns were of one solid casting, necessitating the rejection of the whole apparatus after a comparatively little wear. Nowadays, however, tumblers are almost universally made with wearing plates or shoes and riveted onto the castings, so that they can be removed at intervals and thus the life of the tumbler is much increased. The metal used in these wearing parts is usually nickel steel or manganese steel.

DREDGING MACHINES. 63

As has been mentioned, the lower tumbler is supported on bearings in the casting forming the lower end of the ladder, while the upper tumbler is supported by bearings in a steel bed incorporated into the middle gauntree. An improvement in the lower tumbler is the addition of solid cap-pieces around the ends

Fig. 27. Lower Tumbler on the Ophir.

of the shaft outside the hub. These form grease-cups and prevent the ingress of grit and dirt. Compression grease-cups are used throughout the ladder and bucket-line.

The size of shafting has been necessarily increased with the greater size, capacity, and strength of the modern boats, and upper tumbler-shafts may now be seen of 18 in. diameter.

The bucket-line is undoubtedly the most important, as well as the most expensive, part of a dredge, both in first cost and maintenance. The buckets with their pins and bushings constitute

Fig. 28. Close-Connected Bucket Line on the Butte.

the most vulnerable part of the machine and consequently necessitate more constant repair than any other part. The line may either be open-connected, as in the Risdon type, or close-connected, as in all other designs. In the first case a link is interposed between each bucket and in the second the bucket-

Fig. 29. Open-Connected Bucket Line on the Baggette, showing the New Solid-Forged Links.

bottoms form the links, the rear eye of one being connected to the front eye of the one behind by the pin. The bucket includes a 'bottom' or 'back', a 'hood' and a 'lip', and these are fastened together with a pin and bushings. The bottom is a solid casting of high carbon or nickel steel and very heavy. It takes the wear on the tumblers and when the metal at the eye wears down to breaking thinness it has to be discarded. It weighs from about 500 lb. in the case of a 3 cu. ft. bucket up to 2150 lb. for a 13 cu. ft. bucket.*

The pins are made from a variety of iron and steel, as will be shown later on, and they also vary in diameter; the half-bushings are made almost exclusively of manganese steel. The hoods vary in pattern and are made in several sections riveted together, or in one solid casting or forged piece. The solid cast-steel hood is coming into general use now and will, it is thought, supplant the other sectional rolled types. The lips are an important part of the bucket and as they take the greatest wear they are most often renewed. The parts of modern buckets are designed, however, to out-last each other. The lips are of nickel or manganese steel and in buckets of different capacity they are from 1 to 2½ in. thick, and 8 to 14 in. deep.

Hoods, lips, and bottoms are all riveted together and the capacity of the complete bucket varies from 3 cu. ft. to 13 cu. ft. These are the largest and smallest in use in the Sacramento valley, but the rating is arbitrary and usually refers to the theoretical capacity in wet sand and not to bank measurement, which would be considerably more.

The original idea of an open-connected bucket-line was that it would dig better in hard ground and in large boulders, but the majority of boats, in fact, all modern boats except those of the Risdon type, use the close-connected line in preference; with the latter, more ground can be dug in the same time under the conditions that exist in the districts referred to.

Buckets differ greatly in shape, the chief differences being in length of 'pitch',† angle of lips,‡ comparative depth and width.

The tendency has been to increase the diameter of the pins on the larger boats and they are now made as thick as 4½ in. Some of the pins in the earlier Risdon type boats (still working) are less than 2½ in. diam.; indeed, it is related that the pins used in a

---

* The bucket of this capacity weighs 3200 lb. complete.
† The 'pitch' of a bucket-line is the distance between the centre of any pin and that of the pin in the next adjacent bucket.
‡ This varies with the angle at which the digging ladder is operating and on the amount of wear that has taken place.

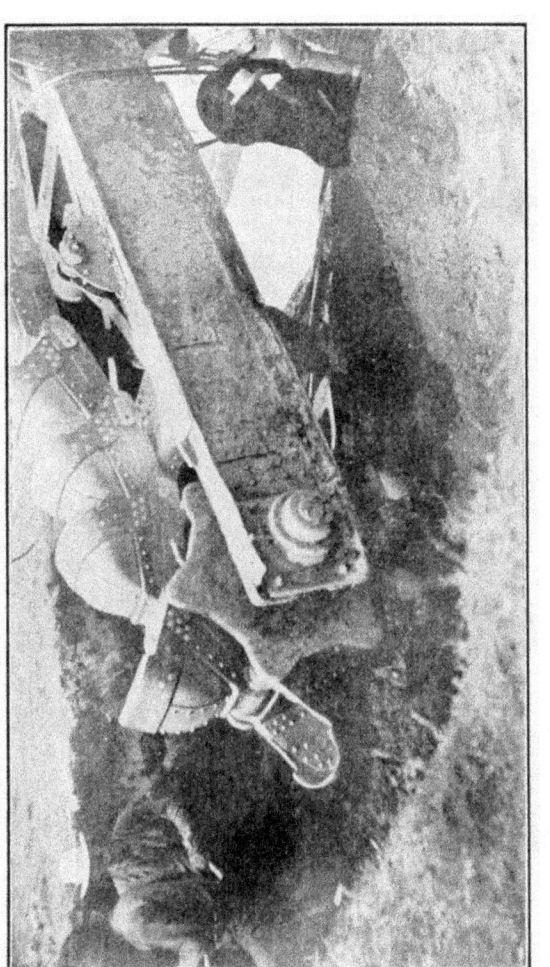

Fig. 30. Repairing Lower Tumbler of Exploration No. 2.

68   DREDGING FOR GOLD IN CALIFORNIA.

bucket-line of one of the first boats built in the State were common 1½-in. carriage bolts and the links were of tire iron ⅜ by 2-in. section. The increase in size, weight, and strength has been extremely rapid and applies to nearly every part of the dredge subjected to wear or strain.

Fig. 31. Bucket of 13 Cubic Feet Capacity, on Folsom No. 4.

The buckets dump over the upper tumbler into a hopper, the steel bottom of which is soon covered with a layer of boulders, preventing the damage that would otherwise result from the continual hammering. On the Yuba boats an idler-wheel is used, as the bucket-line is so long that otherwise it would scrape on the

DREDGING MACHINES. 69

grizzly edge of the well. This arrangement is described in Chapter VI.

Jets of water under head are used to assist the automatic emptying of the buckets as they turn over the upper tumbler and are particularly useful when clay is encountered or in the case of deep narrow buckets, such as the Risdon type. They are arranged in

Fig. 32. Bucket-Emptying Jets on the Butte.

several ways; one, two, or three jets (composed of pipe from 3 to 4 in. diam.) may be placed vertically over one another and be directed at different angles into the bucket or one may be placed at each side, or again, one may be placed at each side and one at the centre; several variations in arrangement may be given by

70 DREDGING FOR GOLD IN CALIFORNIA.

changing the elevations and direction at .which the jets are discharged.

The material is delivered from the hopper to the screen, whose duty it is to disintegrate and classify the material passed through or over it. The two types of screen in use are the flat shaking screen of which, in most boats, two are used (one beyond the other, or one above the other), and the cylindrical revolving trommel, varying in diameter from 3 ft. 6 in. to 6 ft., and in length of actual screening surface from 14 to 21 ft. The largest trommel

Fig. 33. Stacker Discharge on Folsom No. 4.

in use is on the Yuba and is 30 ft. long over all. The dimensions of the shaking screens in use are approximately as follows: Upper screen, from 12 by 4 ft. 6 in. to 16 by 9 ft., and lower screen, from 12 by 5 ft. to 16 ft. by 9 ft. 9 inches.

Probably the largest combined superficial area contained in any shaking screen is afforded by those on the Folsom Development Co's *No. 4* boat, which give 300 sq. ft.. The question of the relative advantages of trommel and shaking screen is discussed in Chapter VI.

The duty of the stacker is to take care of the over-size from the screens and to deliver it at such a point behind the boat that it will not interfere with its navigation.

Fig. 34. Belt-Conveyor on Exploration No. 2.

The stacker is either a belt-conveyor or is of the pan form. The former requires less repair and consequently entails less loss of time—one of the most important items in reducing the cost of dredging; but it usually needs to be replaced oftener and at a large expense. The concensus of opinion among dredge-men connected with the operation of the larger boats favors the belt as a

Fig. 35. Belt-Conveyor Stacker on Folsom No. 4.

method of conveying the gravel from the screens, at least in boats using buckets of 5 cu. ft. capacity and over. The photographs show the two forms of stacker. A pad belt is often used to reduce the wear on the main belt. Motive power for the pans or belt was formerly supplied from the boat, but now in almost every case the motor is placed on the outboard end of the stacker-ladder.

A belt-conveyor may not advantageously be placed at a greater angle than about 19°, while a pan-stacker will carry material at 35°. The length of the stacker-ladder changes with the depth of ground to be dug, as does the digging-ladder, for in deep ground the boat does not move ahead so quickly and a larger amount of material has to be discharged at the stern; therefore to prevent grounding, it must be deposited at a greater distance. To carry the belt over the ladder, 'idlers' are placed at every few feet in sets of three, four, or five, arranged in convex form. The latter is probably the best arrangement as the curve is more even than in the other cases. With three idlers there are two points of more or less abrupt bending, but with four, a V-shape is given to the belt, tending to crack it at the centre. This is more particularly the case in broad and heavy belts and has been found to be the case on the 44-in. belt used on the *No. 4* Folsom Development Co's boat. In one or two cases a practically flat belt with one cylindrical roller, slightly concave at the centre, is used and it is difficult to see why this or, in some cases, a perfectly flat belt is not the best that can be used.* Side boards with canvas attached to them are used and the continual curving and flattening of the belt that takes place at the point of discharge is obviated and the wear is considerably reduced. Besides, the whole width of the belt is utilized as a carrier and the injurious sluicing effect of the water is not so pronounced. In a climate where frost is prevalent, belt-conveyors are not advisable.

The following excellent notes on the electric equipment and transmission are supplied to me by Mr. George L. Holmes.

"The fact that most of the dredges in California and in fact all of the dredges in the Oroville, Yuba, and Folsom districts are driven by electricity has caused their development along different

---

*In this connection Mr. W. P. Hammon says however: "We have just discarded a flat belt-conveyor, because the material spreads to the side board and is retarded by friction, creating an extra load, and further the canvas on the side boards is a continuous source of trouble when digging formation containing clay or mud. Our experience is that the power consumption is increased nearly 50 per cent with this form of belt."

lines than in districts which are not so favored. As a power for dredging purposes, electricity cannot be excelled if proper precautions are taken and the installation is correctly designed. Those, however, who have been using steam-dredges and have remodeled their equipment, installing motors in place of their engines, have come to grief. The reason is plain; when running by steam the boilers, engines, etc., are usually designed so that the engines will stall before the driven parts are strained to a point of rupture. If the dredge-master hangs a monkey-wrench or an old boot filled with gravel on the safety-valve lever and carries a higher pressure than was intended, for a short time, to carry him through a tough piece of digging, he is correspondingly careful while the increased steam-pressure is being carried. Electricity, on the other hand, is brought to the dredge from a power plant generating perhaps thousands of horse-power, over wires of a size sufficient to prevent an excessive 'line loss' and the main motors are, in the latest practice, usually operating under the full voltage of the line. The induction motors in most cases will, before stalling, deliver for a short interval from one and one-half times their rated capacity to twice or three times. Of course precautions are taken, the motors are fused to cut out at 25% over-load over-load circuit-breakers, etc. are installed; in fact, everything is done to prevent an excessive load being transmitted to the dredging machinery. These precautions however really do not prevent entirely the transmission of severe shocks and strains to the machinery, as, for instance, when the buckets strike a large boulder or other comparatively immovable obstacle and are practically stopped instantly thereby, while the motor jumps from its normal power-delivery to two or three times the normal before the fuses or circuit-breakers act. Of course the cut-outs act with 'electrical speed' but even so, there is an interval between the stopping of the bucket and the blowing of the fuse in which mischief may be done. In modern practice, the designer selects the motor which, with his previous knowledge of the art, he believes will be sufficiently powerful to drive the buckets, in the average digging to be encountered in the tract to be worked, at the maxium efficiency of the motor. In heavy ground this motor will, of course, be overloaded and vice versa. The digging apparatus then will be designed to withstand the pull caused by this motor when running at 50% overload and with an ample factor of safety beyond this.

Fig. 36. Belt-Conveyor on Folsom No. 3.

"In the Oroville district we find that the electric power is manipulated in various ways, to serve the required purpose. Some, with a sort of insensate fear of the higher voltages, transform the current on shore to 440 volts, use a ponderous cable for conducting the current on board, and have a comparatively heavy line-loss on account of the low voltage. Others carry the current on board at 4000 volts and transform to 440 for all the motors, while some use the current at 2000 to 2200 volts for the larger motors and transform only for motors under 50 h. p. Each practice has its adherents and its wordy arguments. The best plan, all things considered, is undoubtedly the last mentioned, inasmuch as the losses from line, transformers, etc., are minimized. This plan has been adopted as best by the Yuba and Folsom companies. The usual equipment for a 5 cu. ft. continuous bucket dredge of the type commonly in use at Oroville consists of the following:

75 or 100 h. p. variable speed motor for driving the buckets.
50 h. p. constant speed motor for main pump.
20 h. p. constant speed motor for screens.
20 h. p. constant speed motor for stacker.
5 h. p. constant speed motor for priming pump.
40 h. p. constant speed motor for sand pump.
20 to 25 h. p. variable speed motor for mooring winch.
3 to 30 k. w. O. D. type transformers for the motors under 50 h. p.
1 to 10 k. w. O. D. type transformer for the lighting circuits.
120 to 175 16 C. P. incandesent lamps.

"In the best practice the current is brought aboard the dredge through a cable composed of three strands of insulated conductor well covered and protected with jute and drawn through a rubber hose for further protection against water. Iron-armored cables have been tried but their tendency to kink, break insulation, and burn in two, makes them undesirable. Three terminal panels are provided, one on either side of the rear of the deck-house and one on the bow gauntree, for the connection of the cable when the dredge is working on the spud or on the head-line. From these terminals the current is carried through insulated conductors to the switch-board in the pilot-house. The switch-board may be subdivided into three panels, a receiving panel and two distributing panels. The receiving panel is usually provided with an oil-switch and automatic

Fig. 37. Feather River Exploration No. 1, showing Pan-Stacker.

circuit breaker and with volt and ammeters of the long-scale type. From this panel leads are taken to one distributing panel and to the transformers at the full voltage of the incoming line, from the transformers a lead is carried to the second distributing panel for the lower voltage circuits. Oil-switches are provided for each of the motor circuits, so that each of the lines may be cut out separately in case of accident to the particular motor which it supplies.

"The motors in use in all three of the districts named are all induction motors, the direct current having been found not suitable for the service. There is no doubt that the direct current control for variable speed motors is more efficient and economical than the rheostatic control used on the A. C. motors but the severe service and the liability of stalling a motor at any time, makes the D. C. motor out of the question.

"For the variable-speed dredge-motors, the digging motors principally, use has devised or caused to be devised, a type of rheostatic control commonly known as 'dredge resistance'; it is the ordinary form of grid resistance, but the amount of heat radiating or dissipating surface is proportioned so that the motor may be run continuously on any notch of the controller for a considerable period of time.

"In the variable-speed induction-motors there is of course a great loss of efficiency when running on the lower notches of the controller but, as the power is comparatively cheap, the actual monetary loss is small in proportion to the earnings."

The winchman and controlling mechanism may be situated on the main deck or in a pilot-house on top, preferably on the starboard side, as in this position, with properly arranged windows, he can most conveniently see the bucket-line, upper tumbler, stacker, spuds or head-line, and side-lines, and, at the same time he can manipulate the controlling levers.

The driving motor and pulley, or sprocket, arrangement, are generally placed at the port side forward of the middle gauntree, while the winches are installed opposite on the starboard side of the boat.

Various types of gearing are used in the main drive of the upper tumbler and bucket-line. The typical Risdon design has a single drive-wheel on the starboard side, with the pinion-bearing under-

Fig. 38. Motors and Ladder-Hoist Winch on Boston & California, No. 3.

neath, and resting in a separate casting in the middle gauntree, which in this type of dredge is built of iron. This has been obviated in the most recent machine of this design (See description of *Bagette* boat in Chapter VIII). A sprocket-chain and gear-wheel are used on a few boats, but they do not give the best satisfaction. A somewhat unusual arrangement is the pentagonal equalizing gear used on the *El Oro* boat and shown in Fig. 39, though in the figure the gear-wheel has seven sides instead of five.

Fig. 39. Sprocket-Gear Driving Arrangement on the Pennsylvania.

The equalizing gear is designed to impart a pulsating motion to the driving-sprocket wheel, exactly counteracting the variations in chain-speed above explained. This is accomplished by making the pitch-line of the spur-wheel describe a series of waves, the number of elevations and depressions in which correspond to the number of sprockets on the chain-wheel, and by driving the spur-wheel with an eccentric pinion. The sprocket-wheel and spur-gear are keyed on the head-shaft in proper relative positions. By the use of this gearing less power is required, and the destructive strains due to

Fig. 40. Pentagonal Equalizing Gear.

driving with circular gears are eliminated, thus permitting installations of greater length or the use of lighter chains.

In the usual modern driving arrangements there are two gear-wheels, one on each side of the tumbler (See Fig. 41) and the driving pinions are driven from a counter-shaft by pulley-belt from the motor. Much trouble has been occasioned with winch-clutches and the introduction of a thoroughly satisfactory clutch would be much appreciated. On the *Ashburton* at Folsom a magnetic clutch is used.

Centrifugal pumps are used exclusively on a dredge and include: The main supply pumps of high and low pressure for the gold-saving tables, the screens, and for discharging the buckets;

Fig. 41. Folsom No. 3 Dredge, showing Standard Type of Drive-Gearing.

the sand-pump, used to raise the tailing from the well at the stern of the boat, to which it is conducted when filling up too fast behind; and a small pump for priming the main pump and for general purposes.

The method of mooring and moving the boat is either by spuds or by head-line. The former are long posts, one of which is generally of steel and the other of wood. They are arranged in guides at the stern of the boat and are, as a rule, 24 by 30 in. in

Fig. 42. Wooden Spud of the Ophir.

section and from 50 to 55 in. long. At Folsom a steel spud 36 by 42 in. in section is used. The photographs and sketches accompanying, show their design and construction. The wooden spud is simply a stick of Oregon pine, shod at the lower end with a steel point and fitted with sheave-wheels at the upper end for raising and lowering by rope. The steel spud varies in section but is usually a box-girder constructed of angles and plated with wooden wearing surfaces on each side and often with a web-plate in the centre. It also has a solid steel point and sheave-wheels at the top. In the case of one dredge, the *Pennsylvania,* two round wooden spuds strapped with iron lengthwise are installed, but the boat is now operated by head-line.

Though originally the dredges built by different companies were of distinctive types, modifications have been introduced so rapidly in one and another design that at the present day it is difficult to say that any particular boat represents the special design of a particular company as it would be understood a few years ago. For instance, although dredges built by the Risdon company still retain the open-connected bucket-line, they have dropped the cocoa matting and expanded metal as a gold-saving appliance and adopted riffles and mercury, and in some cases even tables and launders. They have retained, however, in every case, the trommel or revolving screen, and pan-conveyor; in fact, the trommel is being used on many boats of other makes, for example, the *Nevada*, a Bucyrus dredge; and most of the dredges on the Yuba have revolving screens instead of shakers. The machines last mentioned were built and designed by the Bucyrus Co., Marion Steam Shovel Co., and the Boston Shops. The latest of these—*No. 3, 4, 5, 6, 7,* and *8*—are said to belong to the Hammon type of dredge, though they do not radically differ from the Bucyrus design.

At Folsom, shaking screens are used on all the boats except *El Dorado* and at Oroville fifteen boats have shaking screens while sixteen use trommels. The only dredges still retaining the open-connected bucket-line, are eight Risdon boats at Oroville and one at Folsom.* On the Yuba, *No. 1* and *No. 2* dredges have shaking screens, but on all the other boats, revolving trommels are installed.

As there are but two dipper or shovel dredges at present working in the districts under consideration, only a few words will be devoted to them. The *Garden Ranch* dredge, working about four miles south of Oroville, is a fair type of the class and a description of it will suffice. The hull is similar to the ordinary dredge but shorter than the average. The power is electric and only 40 h. p. is required to run it. The digging apparatus consists of a horizontal rotating table at the bow to which is attached an upright crane. This is constructed of two pieces of timber, strengthened and fastened together with iron, between which is fitted the dipper-handle. The dipper or shovel is an iron box open at the top, the

---

*Four Risdon boats at Oroville and one at Bear river have had the bucket lines changed to close-connected buckets, thereby increasing their digging capacity 25 per cent.

DREDGING MACHINES. 85

bottom of which is shut or opened at will by a chain attachment, and the top of the front has a toothed arrangement for digging. The capacity of the dipper is 1⅓ cu. yd. or 36 cu. ft. and the dredge will dig about 20,000 cu. yd. per month. A hopper is attached outside the starboard bow of the dredge at the top, and below this the shaking screen is fitted.

Seven side riffle-tables, pitching back toward the centre of the boat, empty into a longitudinal sluice four feet wide and running

Fig. 43. Broken Steel Spud, Yuba No. 1.

the whole length of the boat; with the tail-sluice emptying at the stern as usual. The stacker is a belt-conveyor running almost flat over the ladder-rollers and has sideboards attached to prevent the belt slipping off the rollers and to retain the material on it. To counteract the weight of the receiving hopper, screens, stacker, and sluices, a hopper filled with rocks is placed on the opposite or port bow side.

The boat is held fast while digging by three spuds, one at the centre of the stern and one at each corner of the bow. These are all of wood, the stern spud being strengthened by iron straps and the bow spud being held from slipping during the dumping of the dipper into the hopper by means of a sort of ratchet-gear. A system of winches and cables is used to operate the crane and dipper and centrifugal pumps supply the water.

The method of operation is as follows: The crane is rotated to the desired spot and the dipper-handle is dropped, by the ratchet-gear, through the crane-beams until the dipper rests on the bottom. The dipper is then caused to scrape along the bottom and up the bank in front, the dipper-handle at the same time sliding back through the crane. The dipper is raised directly over the hopper and dumped by releasing the bottom opening with the chain connection.

The crew consists of: One dredgemaster (acts as blacksmith), 3 lever-men, 3 crane-men, 3 oilers, and 1 laborer. These are employed on three shifts.

The single or double-lift dredge, with extension-sluice, supported by a pontoon, floating at the stern of the dredge-hull proper, has gone entirely out of use in the Sacramento valley, though it is somewhat doubtful whether under proper and careful management it is not in many cases preferable as gold-saver to the present stacker type of dredge. The *Continental* at Oroville, which has since been re-modeled, was a double-lift dredge with an extension-sluice as above. This dredge was built in the spring of 1900. The upper tumbler was comparatively low, and the bucket-line delivered over a grizzly with bars 6 in. apart. The oversize went overboard at a side chute, the fine (under 6 in.) passing through the grizzly and into a trommel with 4-in. holes. The undersize from here (under 4 in.) dropped into a sump in the hull of the boat from where it was raised by a huge centrifugal pump into the long sluice, supported by a pontoon at the rear.

With the single-lift dredge the material after passing over the screen is delivered directly into the long sluice, thus obviating the pumping expense. At Alder Gulch, in Montana, two single-lift sluice dredges, such as above, have been working successfully for several years under the most adverse climatic conditions, and now, a 12 cu. ft. bucket dredge with hull 130 by 48 ft. is being built.

Fig. 44. Well of Yuba No. 8, showing Upper Tumbler, Hopper Grizzly, Idler, and Save-All Sluices before Installation of Bucket-Line.

IV. OPERATION.

The system of operation is as follows: The material is dug by the buckets, elevated and dumped into the hopper at the top, from whence it is fed to the screens. The oversize escapes from the screen and falls into a small chute at the end which delivers it to the conveyor to be carried aft to such a distance that it will not be in the way. The fine material passing through the screen is taken by the water of the washing jets over the gold-saving tables; the

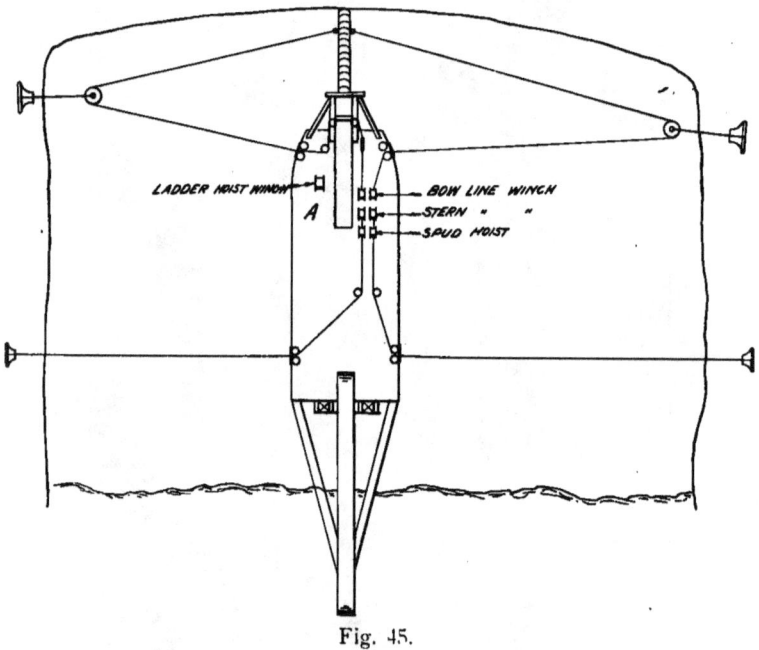

Fig. 45.

tailing is carried by a suitable sluice far enough astern to prevent its interfering with the future operation of the boat. Should it interfere by flowing too close about the stern, the tailing is carried to a well at the end of the hull to be lifted by the sand-pump to the far side of the tailing-pile. As mentioned before, the two methods of working the boats are by spuds and head-line and in the following description it is attempted, assisted by the use of sketches, to make

OPERATION.

clear these methods and the conditions under which each is applicable in both methods of operating. Side-lines are used to move the boat across the pond; these are anchored in the manner shown in Fig. 45 either by 'dead-men', or to any natural object such as a tree, etc., and pass from these anchors through sheaves on the forward deck and on the bucket-ladder of the boat, thence back to the winch-drum. By winding or unwinding the rope on the drum,

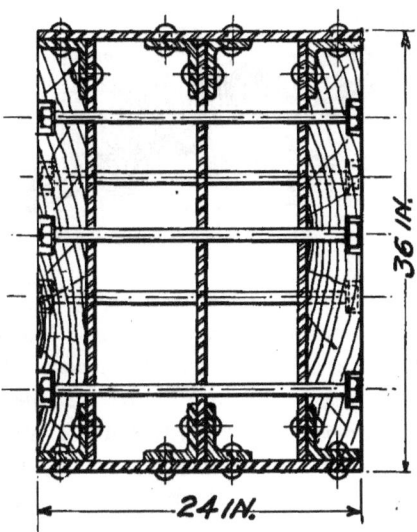

WEB PLATES $\frac{3}{8}$ IN. THICK. END PLATES $\frac{5}{8}$ IN. THICK. ANGLES 4 IN. BY 4 IN. BY $\frac{1}{2}$ IN. THICK.

Fig. 45½. Cross-Section of Steel Spud.

the boat is passed to right or left as desired. In using the 'dead-man' method of anchoring, a T-shaped trench is dug, about 4 ft. deep in that portion which represents the cross of the T and the part that represents the upright of the T slopes from the surface down to the cross part. A piece of wire-rope is firmly attached to a log placed in the bottom of the cross trench and a loop is made at the surface end; to this is attached the land end of the head-line or side-line. Planks are driven down in front of the log and the earth is filled in and tamped; several of these are placed at convenient distances from the pond. Chinese are usually employed for this purpose as well as in removing trees, stumps, etc., and in generally preparing the ground in front of the dredge.

The head-line is attached to the bank similarly to the side-lines but to a different winch-drum on the boat, which is moved backward

Fig. 46. Method of Anchoring a 'Dead-Man'.

and forward by winding and unwinding. The *No. 4* boat of the Folsom Development Co. (13 cu. ft. buckets) has three of these anchorages placed a long distance in front of the boat and the head-line is fastened to each in turn as the boat is moved across the cut by the side-lines. Side-lines vary from ¾ to 1 in. diam. and head-lines from 1¼ to 1½ in. diam. With a head-line, the digging-

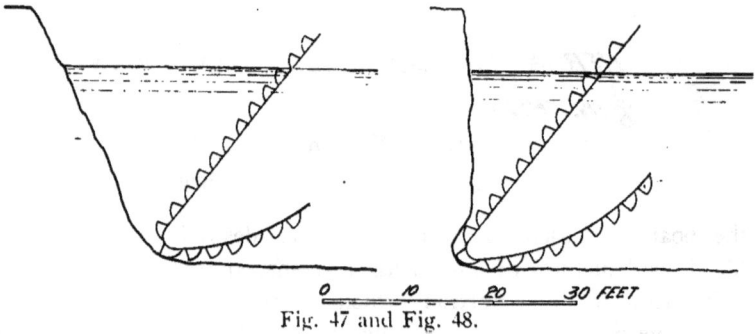

Fig. 47 and Fig. 48.

ladder is kept at the bottom of the digging-face and the tendency is to under-cut the bank and allow it to cave to the buckets, see Fig. 48. This is done by making short vertical cuts of about 5 ft. in height across the bottom of the pond. Some of the material that caves falls over the sides of the buckets, but as these dig and scrape up the

OPERATION. 91

ground for about 20 ft. behind the lower tumblers this caved portion is usually collected. Operation by head-line is best adapted for soft ground.

When the gravel is toughly cemented and the digging is hard, spuds are advisable. The method of using them is as follows:

Fig. 49. Chinaman Removing Trees in Front of a Dredge.

The digging-face of the pit, as in the case of a head-line boat as well, is kept from 150 to 250 ft. wide, and the boat is passed from side to side by the side-lines. The bank is cut in terrace or sloping shape as in Fig. 47. Movement forward and backward is effected by first winding up on the starboard side-line, when the boat will assume the positions shown in Fig. 51. This is called stepping or 'walking' the boat ahead. Care must be exercised in moving the boat with spuds not to place each new pivotal point so that the stacker will cover valuable ground to be dredged on the return cut or fill such portion of the pond as will tend to ground the hull when changing its position. Sometimes the dredging is done in a direction following or crossing a right-of-way stream or an irrigation ditch and the ingenuity of the dredge-master is taxed in keeping the channel open behind him, through, or to one side of his stack-pile.

Fig. 50. Removing Stumps along Irrigation Ditch in Front of the Nevada.

The advantages and disadvantages of spuds as compared to head-lines are as follows: Spuds cause more loss of time in moving, and the stacking problem* is rendered difficult, but harder ground may be dug with less ramming of the bank; the ground may be dug cleaner, and with an uneven surface spuds are preferable. The surging of the head-line and the ramming of the bank are most objectionable features, particularly in hard ground or with open-connected bucket-lines, and there is probably more loss of gold from caved material. On the other hand, the loss of time is less and stacking is easier.

---

*I can see no reason why this should not be facilitated very greatly by using a swinging stacker pivoted at the lower end. This is the method used in sluice practice at Ruby, Montana, and at the Ashburton dredge at Fair Oaks, California.

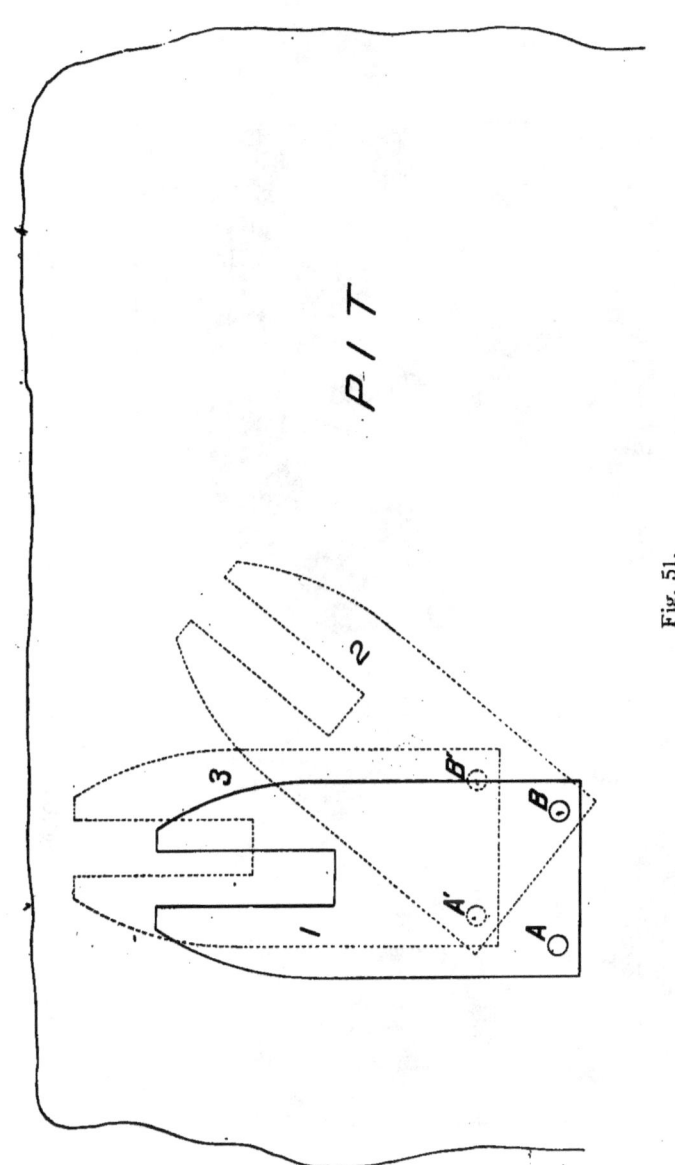

Fig. 51.

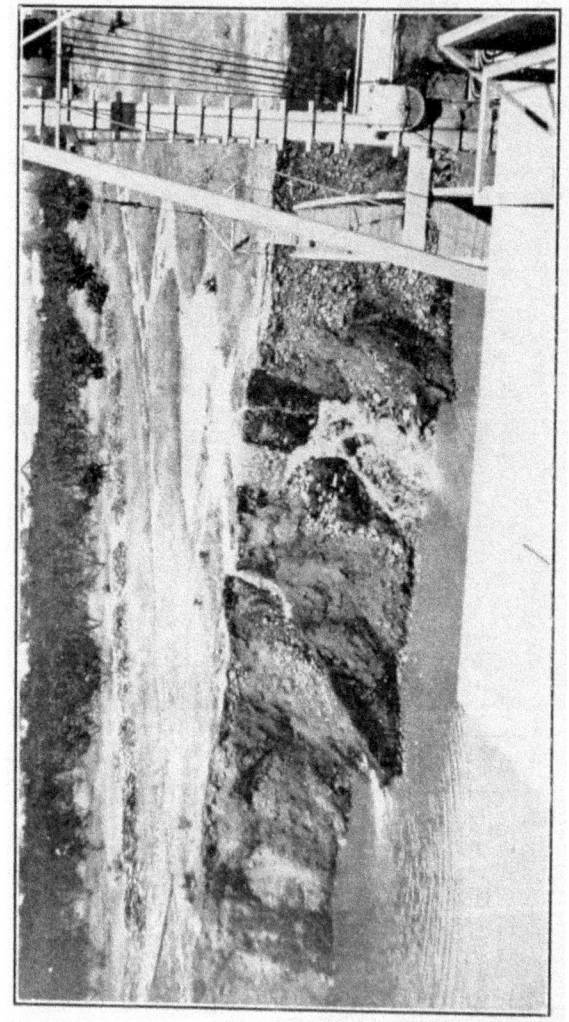

Fig. 52. Working Against Right-of-Way Stream. Orchard Cleared in Front.

OPERATION.

A number of boats nowadays are equipped with both head-line and spud, and either can be used under different conditions. With the use of spuds the plan of the digging-face is in arc-form, the

Fig. 53. Tail Sluices and Spuds on Yuba No. 7. Steel Spud (on the Right) Down; Wooden Spud Up.

boat being swung from one side of the cut to the other in carrying forward the excavation. The whole width of the cut in alternate sections is usually carried forward to the limit of the property and the dredge is turned round and in returning it makes a similar cut alongside the first, but advancing in an opposite direction. When a dredge is completed and ready for work it sometimes happens that it is some distance from its place of operation. In such a case it becomes necessary to dig to the dredging ground. As shallow a trench is dug as will float the hull and occasionally the curious sight is seen of several boats at different elevations climbing or descending a slope formed by the river benches and carrying their own ponds along with them. The water for the ponds is either supplied from the system of irrigation ditches or it is pumped into the pond from the river or some other source. The buckets are designed in shape so that they will best carry the material at the greatest angle at which the ladder digs, which in general practice is not over 45 degrees.

When the pond is below the river, or where the condition is such that the permanent water-level renders the depth too great for the capacity of the dredge, certain expedients must be resorted to. If the gold is at a depth only a few feet greater than the capacity of the dredge, pumping is resorted to and the surface lowered sufficiently to allow the auriferous gravel to be reached by the ladder. Lengthening the ladder and bucket-line is the expedient used on the Yuba and some of these boats are now digging to a depth of 64 ft. below water-level. A somewhat novel and at the same time successful method has been introduced in the case of the *No. 5* boat of the Folsom Development Co. This dredge is working on a high bank several hundred feet above the bench-gravels of the American river and the pay-dirt continues to a depth of 60 ft. below the surface of the ground. By discharging with pumps and (when necessary) filling the ponds by the use of irrigation ditches, the level of the water is kept as nearly as possible at a constant depth of about 35 ft. A bank of about 25 ft. is therefore left in front and above the boat. This is caved by the use of a 'monitor' from the boat supplied by a centrifugal pump with a capacity of 400 gal. per min. (See Fig. 54.) The scheme works well and obviates the long ladder and bucket-line.

The stacker is longer than usual on this boat, but as the sand-pump had to be in use almost constantly, a device consisting of an elevator-belt with small buckets attached was

Fig. 54. Folsom No. 5. Showing Method of Working High Bank with Water Jet.

98   DREDGING FOR GOLD IN CALIFORNIA.

arranged experimentally to lift automatically from the sand-well into the stacker-belt. It was found that so much extra water tended to sluice the material in the belt back into the boat so that now the experimental belt is to be carried up higher, to empty into a launder that will deliver onto the stack-pile. When the gravel proves excessively hard in front of the dredge and is composed of the so-called 'cemented ground', blasting is often employed to facilitate the work. Examples of this work are given below.

Fig. 55. Sand Pump Working on Yuba No. 7.

In one case the dredging pond was 150 ft. wide; at 25 ft. from the face rows of holes were drilled 6 in. diam, at 50 one foot intervals and cased in the usual manner. Cartridges made of sections of 5-in. stove-pipe filled with No. 2 giant powder were used and a cap put on each end and the cylinder covered with gear grease; this was lowered into the holes, the casing withdrawn and the charges fired simultaneously by electricity, 35 lb. of powder being used per hole. In another case, at Oroville, where the depth of dredging ground was

OPERATION.

35 ft., a row of holes was put down at 30 ft. from the face at 40 ft. apart; these were drilled to a depth of 30 ft. each and the charge (100 lb. of 40% dynamite per hole) was inserted in the same manner as above, the blasts being fired by fuse. The cost of this work added 4c. per yd. to the total cost of dredging. In a third instance, the holes were drilled to within 15 ft. of the so-called bedrock (volcanic ash) and charged with 125 lb. of dynamite per hole and fired simultaneously by battery. The extra cost per yard was 2 to 2½ cents.

Most companies have a regular system of laying out their property for dredging and the reports from the foreman, winchman, and superintendent are often elaborate. The Oro Water, Light & Power Co. uses a useful and simple method, the particulars of which were given to me by Karl Krug, engineer and manager of the company. The property to be worked is laid off in 50-ft. squares and stakes are placed at each corner. The lines crossing the property are lettered A, B, C, etc., while those running up and down are numbered 50, 100, 150, etc., and the stakes at the corner are marked to correspond A 50, A 100, B 50, B 100, etc., etc. Spuds are used, and the method followed is to step ahead 10 ft., when the ground is soft and about 7 ft. when hard. A cut 100 ft. wide is carried across in a circular sweep from one to several feet in depth, also depending upon whether the ground is hard or soft. When a distance of 50 ft. ahead has been dug in this way, the boat is backed and moved across and 100 ft. ahead is done, alongside the first cut, the face of the latter then being 50 ft. ahead of the former. The boat is then backed again and moved across to the first cut, and each cut 100 ft. wide is carried on 50 ft. ahead of the one alongside of it, so that when the boundary of the property has been reached a strip 200 ft. wide has been dug the whole length or width as the case may be. Appended are a plan (Fig. 56) and a report, both of which refer to a 24 hr. run of the *Lava Beds No. 3* and show exactly how the records are kept on the dredges of this company.

Referring to these figures and assuming that winchman No. 1 comes on shift at 8:30 a. m., 5 min. are occupied in stepping ahead. During his shift he advances along the zero line of the starboard side of the cut 11 ft. and at the port side along the 150 ft. line, 10 ft. The width of the swing on bedrock is always kept up to the even hundred lines while the excavated surface measurement depends upon the amount of caving. These details are recorded in the small table at the upper left hand corner of the report, as are also the depths and heights of the bank. From these records his cubic yardage per hour may be quickly arrived at. This is generally obtained from the sketch by polar planimeter.

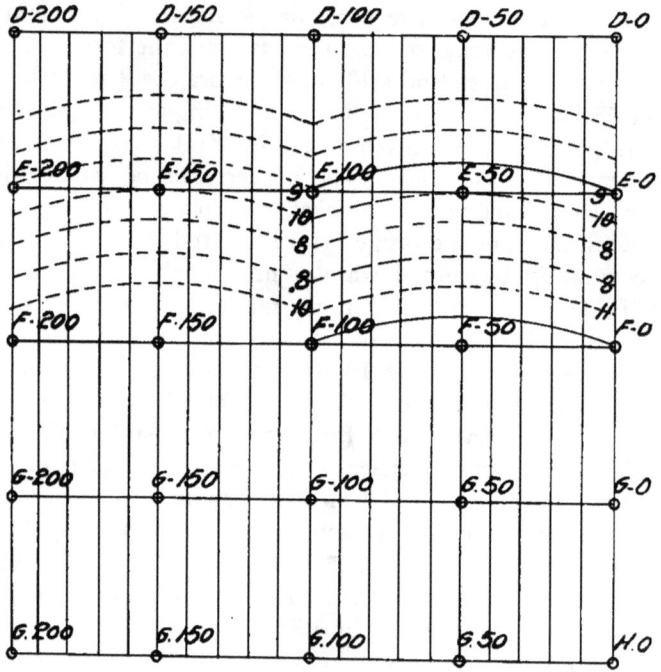

Dotted lines show position of cuts across pit, and prospective cuts

Fig. 56.

Winchman No. 2 comes at 5:28, and No. 3 at 3:06, as shown by the report; their work being shown in a similar manner to No. 1's work. The loss of time and cause of it are also shown on this report and in this case only amounted to 47 min. in the 24 hr.—less than 4%—which of course, is much less than

Fig. 56½.

the average. The winchman's report is checked by the oiler, dredgemaster, master-mechanic, and the manager, before it goes to the head office. As it is well known how much the winchman's capacity is, he has to account for undue lost time in case his work does not come up to the average.

By referring to the records of drill-holes contained in the chapter on Prospecting, it will be seen that the gold lies in a most irregular manner throughout the deposits. It occasionally happens, however, that it is contained over a certain well-defined area in a streak near the bottom or it is definitely known that a certain thickness at the top is barren. In such cases arrangements are made to prevent wear on the tables and sluices by 'blanking' the screens.

On the *Butte,* while changing the boat from one point to another across barren ground where only 'flotation' depth was being dug, unperforated plates were bolted on top of the shaking screens. On the *Pennsylvania,* two steel plates, like hatches, were hinged on each side of the screen, and at another time on the same boat, the material was prevented from delivery into the screen; everything was dumped into the well; but this practice had to be discontinued on account of the rapid filling of the pond. On the *Yuba,* when digging unproductive material, the water was shut off and very little stuff came through the trommels.

Respecting the relative merits of open and close-connected bucket-lines it has been thoroughly proved that at Oroville, Folsom, and on the Yuba, close-connected buckets may be used without any serious hindrance from operating causes, and therefore, it is folly to use an open connection, which may have its advantages in other localities. One of the older boats during the past year at Oroville changed from an open to a close-connected line and increased its monthly yardage from 45,000 to 60,000 cu. yd. per month. Moreover, in practice, other conditions being normal, an open line of buckets will only empty at the rate of from 12 to 13 per min., while a close-connected line empties at the rate of from 19 to 22 per min. As far as the relative filling capacity is concerned, under ordinary conditions and with a proper arrangement, there is apparently little difference between them. A somewhat common accident regarding bucket-lines occurs when they slip over the cheek of the lower tumbler. A method of

OPERATION. 103

replacing the bucket-line when only partly off is as follows: Blocks of wood or sticks of cordwood are inserted between the ladder-frame and the bucket-line just above the cheek of the tumbler on the side on which it happens to be off. The upper tumbler is then reversed and the chain of buckets backed slowly and carefully down. As each piece of wood is carried round the lower tumbler, new pieces are inserted above, and the line is thus gradually worked on. When almost or entirely off the

Fig. 57. Bucket-Line Laid Out on Shore Ready for Installation on the Yuba No. 8.

lower tumbler, the ladder is raised and a piece of steel shafting is poked under the bucket-line and on top of the ladder as close as possible above the cheek of the lower tumbler on the side that the bucket-line is off and in a direction diagonally toward the cheek on the opposite side. It is then lashed to the cheek and the bucket-line backed down as before.

On the Yuba I saw a more extraordinary case than either of those just explained. At *No. 2* dredge for some unaccountable reason, after the bucket-line came off the lower tumbler, winching

104   DREDGING FOR GOLD IN CALIFORNIA.

was continued for some time and two of the ladder-hangers were carried away before the motive power was shut off. By that

Fig. 58.   Accident to Bucket-Line on Yuba No. 2.

time the bucket-line was hanging over one side of the ladder and in the well of the boat. The line was uncoupled as near the lower tumbler as possible and the lower end made fast to a

OPERATION.

crucible-steel cable 1¼ in. diam., kept for just such an occasion. The line was rotated in the same direction as when digging, the lower end at the same time being drawn out on the bank. On emerging from the water a most extraordinary condition was observed. The heavy chain of ungainly buckets, each weighing probably over 1500 lb., was twisted into a perfect loop, the actual circle or loop being completed in eight buckets. Fourteen buckets containing the offending loop were detached and hung up as shown in Fig 58, and after taking them apart

Fig. 59. Cable Transport by Board Trestles.

the line was straightened out on the bank, put together, and made fast to the portion on the ladder. The upper tumbler was then rotated and the line coupled. The accident happened at 6:30 on Friday, June 8, and the boat resumed digging on the following Monday afternoon.

A more important consideration than one might suppose is the necessity for keeping the boats scrupulously clean. This is applicable, of course, to all machinery, but there is so much mud, water, and grease in the confined space upon a dredge, that it becomes a most important factor in its life and serviceable work.

Besides, it costs little to effect this. Two of the boats in particular at Oroville are noticeable for this characteristic and both

Fig. 60. Cable Transport by Barrel Pontoons.

the boats and parts are kept constantly washed down with a hose and at intervals they are painted, etc. Care is taken to cover grease-leaks so that the amalgamation will not be effected and

OPERATION.

receptacles are placed to catch the drip, etc. On one of the Yuba boats the same care is conspicuous and a large sign hung in a prominent place carries out the idea:

> OUR MOTTO:
>
> A PLACE FOR EVERYTHING AND
>
> EVERYTHING IN ITS PLACE.

The contrast, however, between this boat and some of its neigh-

Fig. 61. Cable Transport by Forward Gauntree.

bors was partly due to the small amount of clay that it had to contend with.

The insulated cables that transmit the electric current to the dredges from the sub-stations are generally of the submarine armored type and as immersion in water for any length of time is bound to have, either directly or indirectly, an injurious effect upon these, several methods are employed by which they are conveyed across the ponds. At the *Nevada,* ordinary floats made of boards in the shape of small cross-trees are used while at the *Butte* pontoons made of barrels are employed and appear to answer the purpose well. The *Leggett No. 3* suspends its cable by means of a hanger from the forward gauntree frame but it is contended that the jarring from the digging-ladder is transmitted more directly to the cable in this manner and has an injurious effect on it. The best method that I observed is that now used at Folsom, where the cable is passed through rubber hose and effectually protected from water.

V. THE METALLURGY OF DREDGING.

The gold-saving appliances of a dredge consist respectively of screens, tables, and sluices.

The essential duty of the screen is to classify the material prior to concentration, it also serves to disintegrate or break up the material passing over or through it, so that particles of gold may not be carried off in lumps of clay or cemented gravel, to be lost by passing out at the lower end over the stacker. The aim also, of course, is to prevent the larger gravel and boulders from being washed over the sluices. It is held by some authorities that the old fashioned single-lift dredge, where everything passed from the upper tumbler over long sluice-boxes with riffles supported on a pontoon behind the dredge, is of equal gold-saving efficiency and more economical (when properly designed and managed) than the present style. There are cases too where the double-lift dredge with the long sluice might be advantageously replaced by the single-lift pattern. As to the choice between the revolving trommel and the shaking screen, those who use the former claim that when constructed, as it usually is, with flanges and rods across it, it turns over lumps and exposes them to the action of the water-jets on all sides. On the contrary, those who advocate the shaking screen contend that the jets have a thinner surface to play upon, the actual screening is larger and the material is deposited on a wider surface. From the modern standpoint, however, the last consideration does not count for much.

The first cost and repairs of the revolving screen are usually greater than those of the shaking type. The diameter of the holes is governed by the size of the gold; as this is usually very fine throughout the Sacramento valley, the holes vary from $\frac{5}{16}$ in. to $\frac{1}{2}$ in. at the upper part of the screen and from $\frac{1}{2}$ to $\frac{5}{8}$ in. at the lower end. The holes are reamed on the outer side to give a free discharge. Anyone who has watched the action of either sort of screen for any length of time, particularly on boats where a sticky clay is treated, does not have to be told that neither of them is a thoroughly efficient machine for the work. Quantities of fine material may be seen passing to the stack-piles.

The patterns of the gold-saving tables and sluices differ widely. Formerly, on the Risdon boats, cocoa matting with expanded metal in diamond-mesh pattern was used, but this material has been almost exclusively replaced by riffles and quicksilver, but the designers still use the tables sloping toward each side of the boat at right angles to the slope of the screen. In boats of modern design various devices are used under the screens.

Fig. 62. Tail Sluice, showing both Angle-Iron Riffles and Cocoa Matting with Expanded Metal.

A certain proportion of the material delivered by the bucket over the upper tumbler does not go into the screen-hopper, being carried too far forward in its delivery. A grizzly is therefore arranged at the edge of the well; the undersize falls onto a short sluice with riffles (sometimes two are used, one under the other and sloping in opposite directions), which either empties directly into the well or (when sloping toward the stern) delivers into a pipe passing through the deck, or it may be carried down the entire length under the screen to the stern of the boat. On the Yuba boats where deep gravel is encountered and long ladders

Fig. 63. Trommel, Pan-Stacker, Side Tables, and Stream-Down Sluice on Exploration No. 1.

are used, the tendency of the sag of the bucket-chain is to scrape against the edge of the well. An idler (a large flanged wheel or drum), is installed beneath the upper tumbler and serves to take up this sag. Although 'save-all' sluices have been placed on these boats, they are not of much service. In most cases on the Yuba the riffles are either banked with sand or carried away altogether by falling boulders from the overflow of the screen-

Fig. 64. Clean-Up Apparatus and Riffles on the Butte. (a) Pot. (b) Strainer. (c) Scoop. (d) Pan. (e) Riffles and Mercury Surface. (f) Appel's Mercury Surface.

hopper. This is to be obviated by placing riffle-bars between the flanges of the idler-drum. Various gold-saving devices are used on the tables and sluices, but the angle-iron riffle is the most popular. This may be placed across the sluice-box or lengthwise, with small stones between the bars; a combination of these two is often used. Quicksilver, of course, is placed behind the riffles and at intervals, near the head of the tables there is a mercury trap of variable design. The best I saw and that giving the largest and cleanest surface was one designed by Mr. Harrison

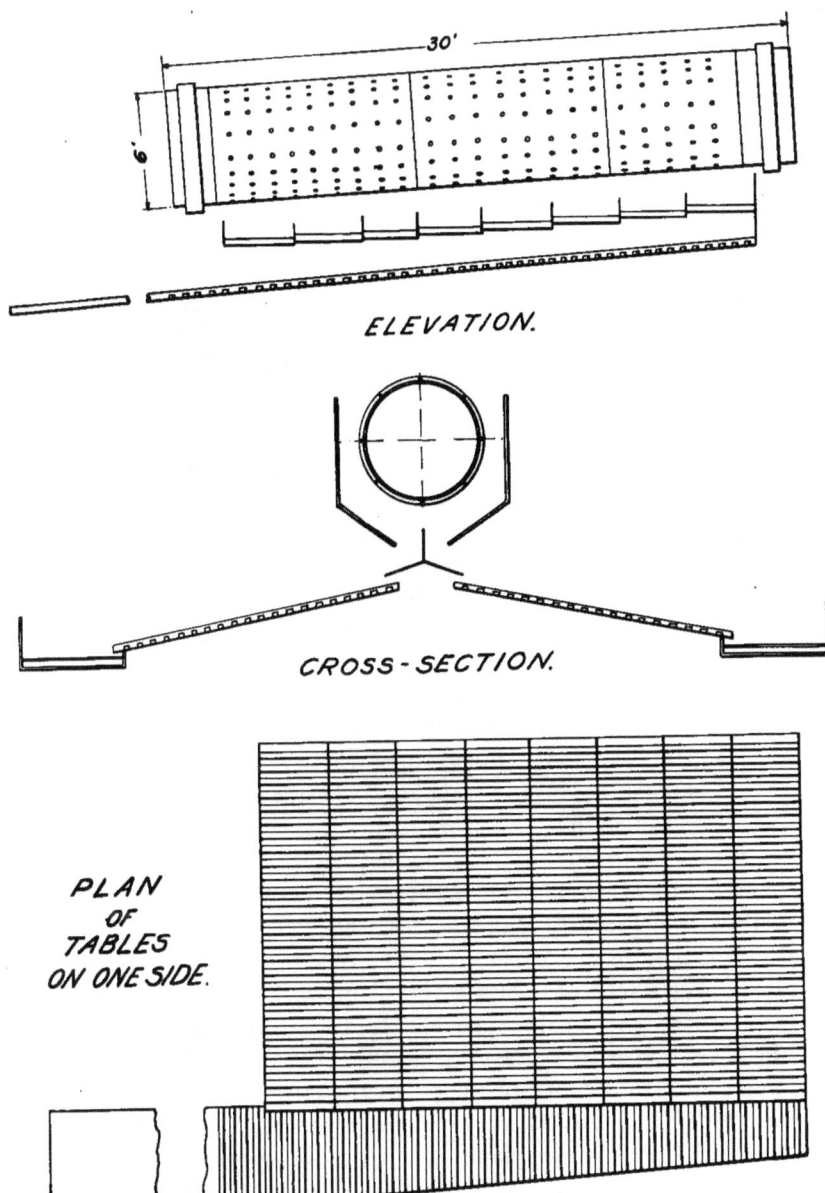

Fig. 65.

Appel, at Oroville, and used on the *Butte* and *El Oro* dredges. It is made of wood with thin strips of metal on edge placed at intervals of ½ in. apart. Small blocks of wood keep the strips separate.

The chief value of cocoa matting is as a collector of fine gold, but the accumulation of sand and slime interferes, so that in the clean-up often only the coarse gold that settles behind the metal is recovered. It is a good idea to use a piece of this material near the foot of the plates, and indeed, some experiments are being made in this direction at the present time.

To illustrate some of the different screening and gold-saving arrangements on the various boats, the following specific cases are described by means of rough sketches. As an example of the simpler modern boat, the *Yuba No. 4* has the arrangement shown in Fig. 65. Eight tables sloping at right angles to the revolving screen empty into a longitudinal sluice called the stream-down box; this is parallel to the screen. The tables and sluice are fitted with wooden riffles topped with iron bands, placed as shown in detail sketches. The bars are 1 in. high and 1 in. wide, and the spaces between are 1 in. wide; the tables are 30 in. wide and the 'stream-down sluice' is 18 in. wide at top, 4 ft. 3 in. wide at the tail and 30 ft. long. A plain iron sluice extension without riffles carries off the tailing.

A similar arrangement is used on the *Boston & California No. 3* and other boats. More complicated schemes are employed on the *Pennsylvania* at Oroville, built by the Miners' Iron Works, of San Francisco, and the *Biggs No. 2* (now *Exploration No. 2*) built by the Bucyrus company. In the first case two flat shaking-screens are installed one over the other, with opposite motion. The upper screen has holes ½ in. diam., and the holes in the lower screen are $\frac{5}{16}$ in. at the upper end and $\frac{3}{8}$ in. at the lower. Under the lower screen are six layers of plank (ten in each of the two upper layers and nine in each of the others) placed across the direction of the slope of screens, at intervals, the layers sloping alternately forward and aft. These planks are grooved lengthwise with blocks placed in the grooves at intervals; they break joint in alternate grooves and are drawn out at clean-up. This forms a mercury trough. Under the layers of plank are ten tables sloping at right angles to slope of screens, each about 18 in. wide and fitted with amalgamating surfaces, planks are placed cross-wise with 1 in. diam. holes bored in them, cocoa matting, and cross-riffles. These empty into a 'stream-down box' with cross-riffles and end-riffles at

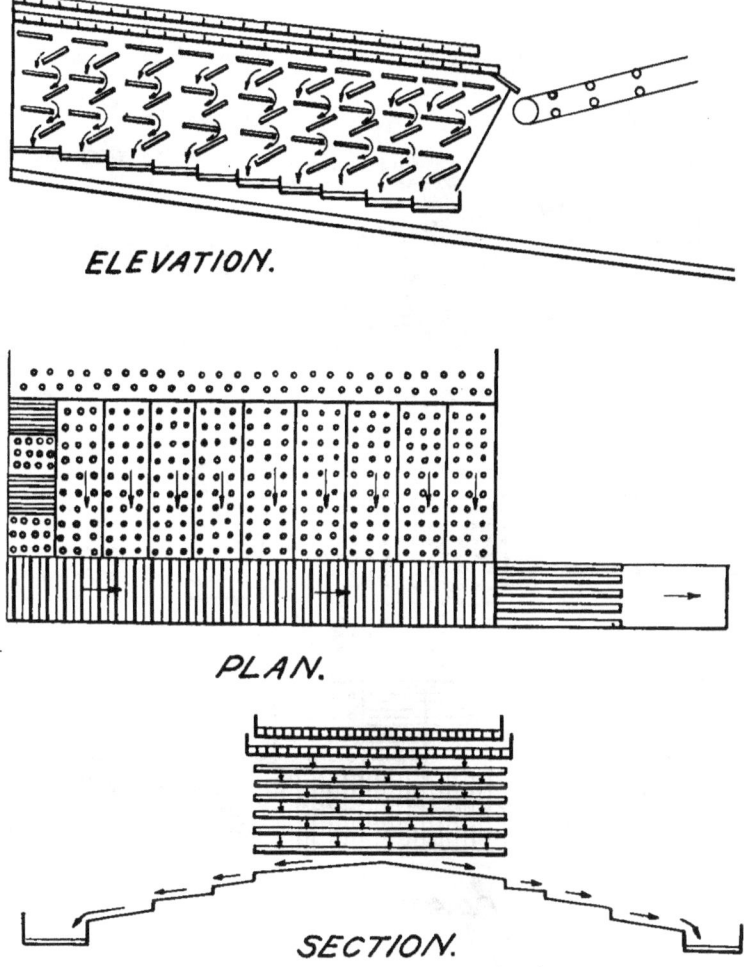

Fig. 66. Tables on the Pennsylvania.

the lower sections. The side tables have two drops between the screen and stream-down sluice.

The *Biggs No. 2* has a single shaking screen. Under this are two layers of plank placed cross-wise and extending about

116   DREDGING FOR GOLD IN CALIFORNIA.

half way from the top of the screen toward the centre. The majority of these are fitted with small cross-riffles. Below (See Fig. 67) are the riffle tables and stream-down box, similar to those on the *Yuba No. 4* (See Fig. 65).

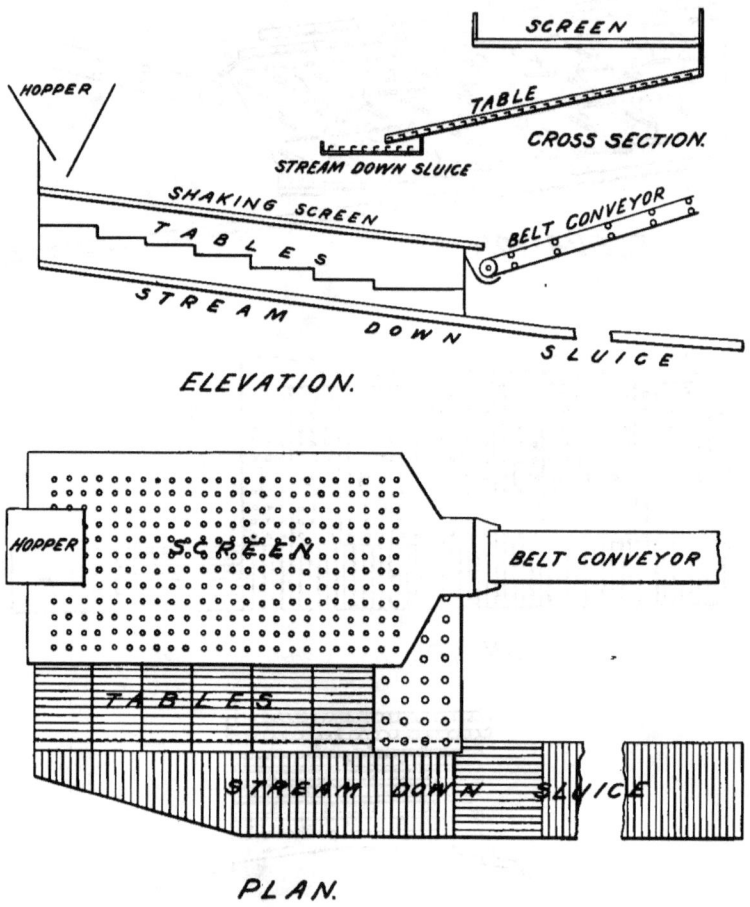

Fig. 67.

On the *Leggett No. 3*, a Risdon boat built in 1904 at Oroville, the arrangement is somewhat different (See Fig. 68). A revolving trommel 54 in. diam. is used, built in four sections lengthwise. The total length of screening surface is 14 ft. The holes in the three upper sections are ⅜ in. diam. and the lower

THE METALLURGY OF DREDGING. 117

section has ½ in. holes. There is one row of square holes two inches around the top, to allow small stones to come through for filling riffle-spaces. The purpose of these stones is to protrude above the riffles and stir up the sand as it passes,

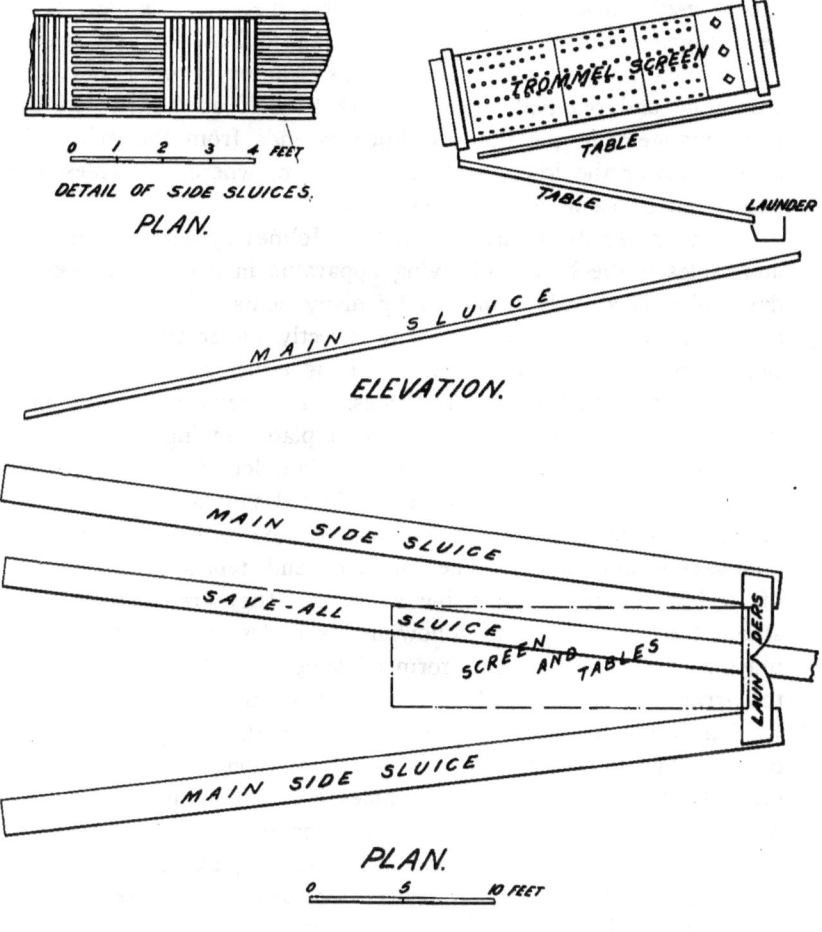

Fig. 68.

and they appear to take up natural positions in the riffles better than if placed there by hand. This allows the gold particles to sink into the quicksilver. The two tables under the screen feed to two launders, which deliver in turn to the main long sluices on each side and which, with the tables under the screen, are arranged with sections of angle-iron riffles, most of them placed end-ways, as shown in sketch. The save-all sluice extends from under the screen-launders to the stern of the boat under the screen-tables. The side sluices are now each 2 ft. wide and the save-all sluice is 18 in. wide. This method, however, is to be changed and simplified and, according to Mr. James Leggett, made much more efficient. The time of clean-up is to be shortened from one and one-half or two hours to half an hour. In the new arrangement, the save-all sluice extends from the grizzly to a point under the lower end of the screen, where it divides and passes out on each side of the stacker.

It is generally admitted that the Holmes system of launders and tables is the best gold-saving apparatus in use at the present day and it is already employed by many boats. They consist of a wide table or tray of iron plate directly under the screen and sloping in the same direction. This is either fitted with riffles or is a plain steel plate with sides. It receives the screening that empties from the end onto another plate, sloping in the opposite direction; this in turn delivers into launders (Fig. 69), which are divided, and empty onto a set of divided sluices sloping toward the stern of the boat on either side and fitted with riffles or other gold-saving apparatus. The launders and tables are generally of steel with steel bottoms, but sometimes they are partially constructed of wood. The following examples will illustrate the practical application of this form of table: In Fig. 70 is shown the arrangement employed on the *Baggette*, the latest and largest Risdon boat, the trommel screen is 29 ft. 6 in. long and 6 ft. diam. The screening surface is 21 ft. long; this length is divided into five sections. The holes in the two upper divisions are $\frac{5}{16}$ in. diam. and in the three lower portions are $\frac{3}{8}$ in. Directly underneath the screen is a plate 8 ft., as long as the screen and fitted with angle-rich cross-riffles. Two lower plates are also arranged with riffles. They empty into the launders as shown in Fig. 69. The launders are of iron plate, with double bottoms and deliver to the four sluices on each side, each 30 in. wide and 12 ft. long. The 'save-all' sluice is 30 in. wide by 28 ft. long.

On the *Butte*, the arrangement is slightly different. Two shaking screens are employed, the upper one 13 ft. long by 4 ft.

Fig. 69. Screen, Table, Launders, and Sluices (Holmes) in Course of Construction on El Oro Dredge.

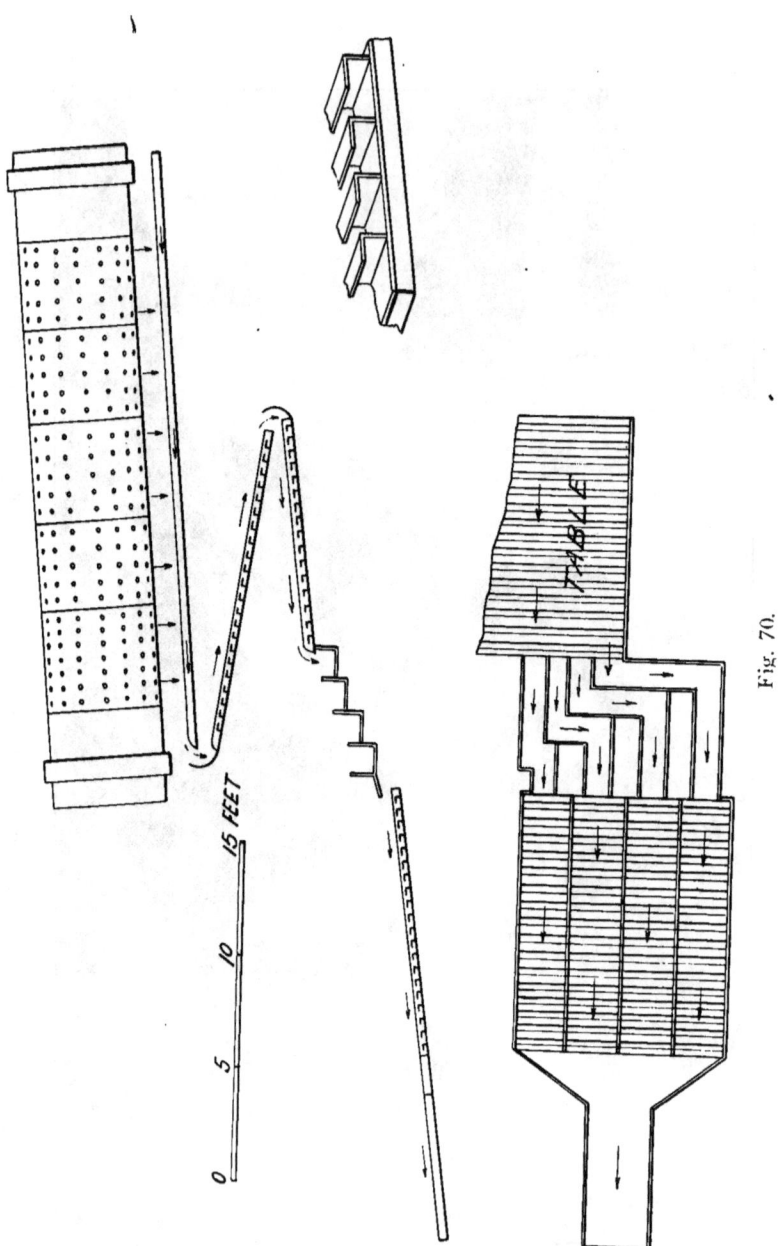

Fig. 70.

Fig. 71. Launder Delivery and Lower Tables, with Angle-Iron Riffles, on the Baggette.

Fig. 72. Tail Sluice on the Baggette.

8¼ in. wide, and the lower one 13 ft. long by 5 ft. 7 in. wide and punctured with ⅜-in. holes; the undersize is carried onto a steel tray divided lengthwise by three angle-irons and sloping with the screen. These irons prevent accumulation of sand when the boat

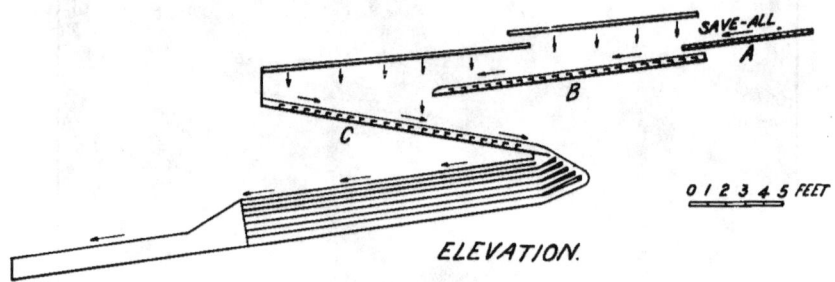

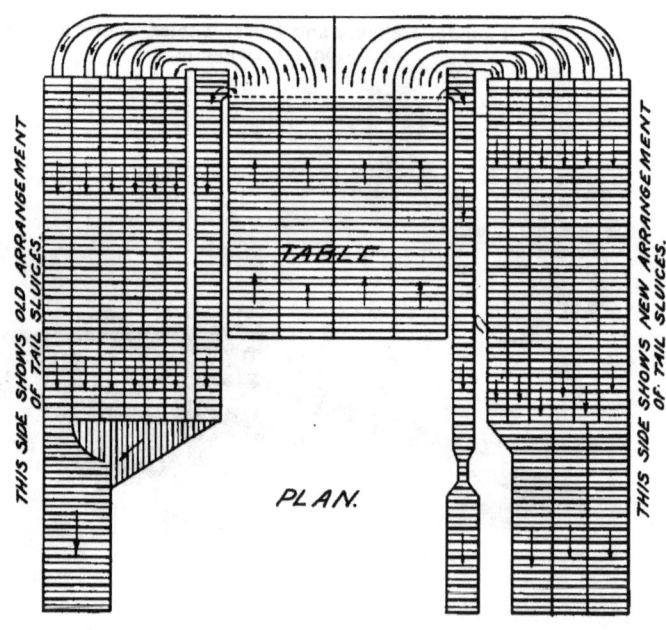

Fig. 73.

lists. From the lower end of this tray the flow is onto a riffle-plate, B, the same width, which in turn delivers into the launders, and thence to the sluice-plates.

THE METALLURGY OF DREDGING. 123

The arrangements on *El Oro* are practically similar, with some slight change in the plan of tail-sluice.

A brief description of the gold-saving apparatus on the *Folsom No. 4* will be interesting, as it is the largest dredge in the world. Double shaking screens are used, the upper one 10 ft. 6 in. long by 11 ft. 11 in. wide and the lower one 14 ft. long by 12 ft. 8 in. wide. The save-all sluice *A,* (Fig. 73) is at such a height that it delivers onto plate *B,* under the upper shaking screen. Table *B* delivers onto *C* and *C* into the launders, which are curved

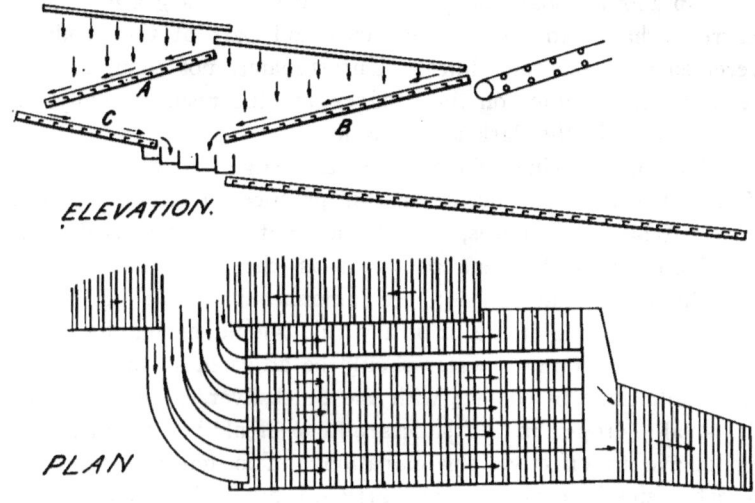

Fig. 74.

and drop in steps toward the outside of the boat corresponding to the steps in the sluices, of which there are seven, running lengthwise with the boat. Cross-riffles are used and stops are fitted at the points shown.

No. 5 boat at Folsom has a most interesting arrangement, recently installed. The screens are not changed; the dimensions are: Upper screen 16 ft. long by 11 ft. 10 in. wide; lower screen 16 ft. long by 8 ft. 9 in. wide. They were arranged originally as in Fig. 74. The undersize from the upper screen dropped onto table *A* and that from the lower onto table *B.* Table *A* delivered onto plate *C,* and *C* and *B* delivered into the launders, which were situated about midway under the screens. Plates *A*, *B*, and *C* were all riffled. The launders fed onto a series of Holmes tables, as shown in plan. In the new arrangement, table *C* is removed altogether. Table *A* receives the undersize from the upper screen and delivers to one set of launders and onto three tables, which have been raised up. Plate *B* delivers the undersize from the lower screen into a set of launders 4 ft. lower than the first set and these deliver in turn onto another and lower set of tables. Thus the product is divided and the separate clean-up becomes useful in testing.

On another boat belonging to this company, a grating (undercurrent) has been put in the sluice and part of the product is received on a separate sluice. Unfortunately not enough of such testing work is done on the dredges and the results usually leave the operator in the dark as to his losses.

The gold-saving efficiency of a dredge is affected by two factors that also influence stamp-mill practice. I refer to the grade of the tables and sluices, and the amount of water used. It is considered that $1\frac{1}{2}$ in. per foot is the most efficient slope for the tables and sluices; the amount of water is variable. As a matter of fact, no arbitrary rule should be blindly applied. The slope and flow of water should be adjusted so that all of the gold will come in contact with the quicksilver. Impact, assisted by small 'drops' in the sluices and a break in the current are conducive to effective amalgamation. Too much water at too great a speed, however, may carry off fine gold, some of which in the Oroville deposits, is (to use a well-worn illustration) so fine as not to settle after two hours in a bottle containing a weak acid solution. On the other hand, if the grade is so flat or the volume of water so low as to allow an accumulation of sand to cover the mercury, amalgamation is hindered. The filling of the interstices of the cocoa matting by fine sand or slime is a drawback to that method of gold saving.

It often happens that the boat has a list according to the pull on side lines, and the sluices on one side may be seen banked with sand, while the riffles are covered. Almost every dredge has a variable 'flotation' level at either end, depending on the hardness of the ground being dredged, and the depth to which the ladder is digging. Due to this cause and the list, therefore,

Fig. 75. Stacker and Remodeled Sluices on Folsom No. 4.

it is seldom that the sluices and tables maintain the most favorable angle. In any case, the gold is so excessively fine, particularly at Oroville, that much of it is lost even under the most favorable conditions possible with the present gold-saving devices. The extent of this loss is not known, as the small amount of research work has given no results of any value. This is one of the features of dredging that surprises the millman and metallurgist. Even the technical mining engineer whose chief work lies among placers—and this is now recognized (particularly in the West) as one of the specialties of the profession—has little to offer in the way of authentic data regarding the actual contents of the ground that he may be exploiting and consequently he cannot know his losses. No proper testing of the tailing has been made; automatic sampling is unknown. Nevertheless, on being questioned, the manager will tell you that he has sampled his tailing and knows just what his losses are. On closer interrogation he says that he has made his conclusions by careful pan and rocker work at the tail-sluice and that what a pan or rocker will not save, no sluice or table will. At assays, he sneers.

Many mills built ten years ago are now known to have lost large sums in gold and silver that might have been recoverable; how many mines are now, by improved appliances in the mill proper, by fine grinding, cyanide, chlorination, or other chemical or mechanical means, reducing ore at a profit that ten years ago was pronounced by engineers as 'too low-grade' to pay? Consider the comparatively recent improvements on the Rand, in Australia, and elsewhere in slime-separation and the consequent increased recovery. In some cases dividends are being paid entirely from this saving.

The dredge operator does not care what the ground may contain—by assay—so long as he is sure that with the present devices he is saving all that is commercially possible. At the same time he is open to consider any improved methods of saving you may offer and says, in effect, that when the process has been proved practicable, he will examine his ground (which will no doubt then be worked out) to see if it contains the value which your method may recover! Not very logical, to say the least of it. What would have been the present state of the industry if

THE METALLURGY OF DREDGING. 127

this had been the standpoint of miners since the days of which Pliny writes?

There is certainly a large loss, due to mechanical arrangements, in the material that passes over the upper tumbler and into the well. This is increased on the Yuba boats, where the idler arrangement makes it difficult to take proper advantage of the save-all. Even in boats without this idler, the loss in the well is known to be appreciable. It is not of much use to attempt to find an accurate method of testing the gravel in place before dredging, as the point to be decided is the absolute amount of gold that is actually passing through the dredge and being lost at the tail (including that which goes over the stacker), and all the gravel is not passed through the dredge.

Even the character of this gold is not known except that some of it is excessively fine. It is possible that a proportion of it is not only fine, but coated or otherwise unfit for amalgamation. 'Rusty' gold, so-called, due to a coating of silica and sesqui-oxide of iron is not at all amalgamable and what little of it is caught is due entirely to its specific gravity. The results of an interesting experiment conducted by Mr. Charles Helman, partly in my presence, are indicative of possibilities in the direction of such research. In the first place, however, it should be said that accurate automatic sampling must be applied over a proper period to get correct results. No dipping or digging of a haphazard sample from the tailing-bed will do, nor is the method of placing a tub at the end of the sluice efficient, for the tendency of the finest silt (probably the most valuable portion) is to flow off in the rush of water. In the case mentioned above, some of the muddy water collected carelessly from one or two buckets of the sand at a clean-up was filtered and the resultant impalpable reddish powder assayed $9600 per ton in gold and not only could not a 'color' be detected with the strongest glass but it was also found that none of it would amalgamate! This powder, it is true, represented a certain amount of concentration, but it would stand a great deal of dilution, and still carry values that would make even a chemical method of treatment profitable. With dredges of the single-lift and long-sluice pattern the rehandling of this material would be facilitated. The finer portion of the tailing might be treated on the bank, if valuable enough.

An instance of testing the tailing at the Conrey placer mines at Alder Gulch, Montana, is of more than passing interest, most work of this description having been of a very perfunctory nature. Single-lift dredges with long sluices supported on a pontoon are used in this district and holes were drilled through the tailing behind the dredge with a Keystone machine. The material was found to be unproductive, with the exception of the bottom layer of 6 in. or 1 ft., where gold values were recovered that made an average of 4 to 6c. for the whole tailing mass.

An extract from one of the earlier reports of the California Bureau of Mines describes tests of muddy water from some of the best of the Grass Valley mills. In 12 tests, where water was taken from a point nearly a mile below the mills, returns were obtained by assay of about $2 per ton. It was estimated that 576,000 gal. of this water flowed every 24 hours, which meant $339 of float gold, to say nothing of loss by imperfect pulverization. A clean-up of the twentieth undercurrent of the Spring Valley Hydraulic Co., at Cherokee, Butte County, Cal. yielded $2600. This gold was taken from material that had been sluiced over 2½ miles of sluices and over 19 other undercurrents, and even then Chinese were working the tailing below this point; moreover, the gold-saving system of this company was supposed to be the most complete in the State, so that it may be easily seen how loss occurs over the short-sluice system of a dredge.

The clean-up is made at intervals that vary on the different dredges, as also the time it takes to accomplish it. As extraction ceases during the clean-up, it is important that it should not be done oftener than is necessary to prevent loss by overloading the tables and sluices. In some cases certain tables are shut off from the screen-delivery, the flow being allowed to continue over the others; in this case digging is not interrupted. It is doubtful, however, if this is good practice. In other cases, the two sides are cleaned up on alternate days, digging continuing all the time. Usually a special set of men is kept for this purpose where several boats are in operation. On the Yuba until recently there were six boats operated by one company and in each case one day was devoted to the clean-up. This work is done by three

Fig. 76. Trommel on Yuba No. 8, 30 Feet Long and 6 Feet Diameter.

men who also measure the bank, do surveying and other work, and when necessary, call on the boat's crew for assistance. One of the party attends to the melting. In this way the men become expert and time is saved. The enforced idleness during clean-up time is employed in making repairs, oiling, etc. Of course, in the case of a company with only one or two boats, it would not be practicable to employ a special clean-up crew.

To illustrate the most general practice in cleaning up, a description of the process on *Yuba No. 4* is given. A revolving trommel delivers onto eight tables on each side and these empty into a stream-down sluice (See Fig. 65). All of the tables are fitted with wooden riffles capped with a flat iron band. These are 1 in. square in section and placed $1\frac{1}{4}$ in. apart. The side tables are 30 in. wide and 13 ft. long, and are fitted with two stops or permanent bars, one at centre and one at the lower end. The stream-down sluice is 18 in. wide at the upper end, 4 ft. 3 in. wide at the lower end and 30 ft. long. The clean-up is done on both sides of the boat at the same time, by the clean-up force assisted by the regular crew. The screen-motor is stopped and, of course, digging ceases. Two men loosen up the riffles on the upper side-table with a bar and these are washed onto the table with a hose and laid aside. The hose is then used—always pointed up-sluice to wash down the coarse gravel—and each man taking a section above the stops, stirs the material with a small rake, something like a currycomb, so that the coarse material passes over the stop. The amalgam, quicksilver, and black sand remain behind the stops; the clean-up men pass onto the next table, and so on, until all of the side-tables are finished in the same manner. Meanwhile two other men with buckets and a wooden hand-trowel scrape the material—amalgam, quicksilver, and black sand—into the buckets, which are then emptied into a wooden tank that is kept full of water. When the tables have been finished, the water is removed from the tank (with a basin) and the material is lifted (with a scoop) and carefully fed into the hopper of a 'long tom' constructed as shown in Fig. 77. This sluice is 14 ft. long and the width inside the box is $12\frac{1}{2}$ inches.

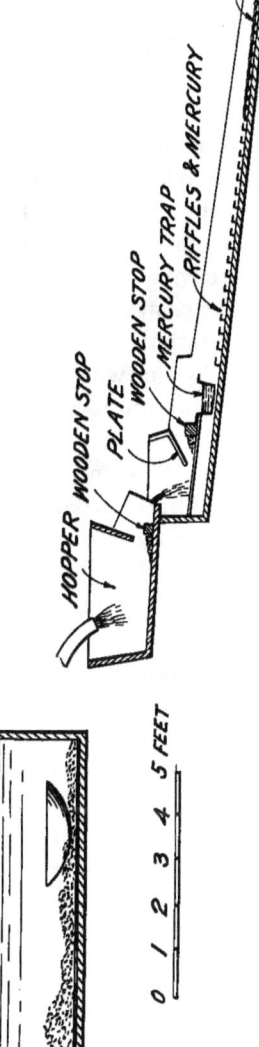

Fig. 77

Water is fed into the hopper by a hose from the pump. The mercury trap is kept stirred with a long spike and 90% of the amalgam is caught above the stop B just above the mercury trap. The amalgam is removed from the sluice and drained in a small inclined metal trough and the 'quick' is collected from the trap' in the long tom and strained. The sand from the side-tables is put back into the vat and passed through the 'tom' again with the material collected from the 'stream-down sluice. The amalgam and mercury are thus collected, while the sand remaining at the end of the process is replaced on the sluices. A 'long tom' is used on each side of the boat and the results from the side-tables are mixed together for retorting and melting, as is the material from the stream-down sluices on each side. The trommel rotates from starboard to port, like the hands of a clock and the material is usually thrown up on the port side of the screen, but with the side-tables the screen-delivery is arranged so that most of it goes to the starboard tables and most of the gold is caught on this side. The clean-up commences at 7 a. m. and is completed in from 2 to 3 hr. Minor repairs are usually done at this time and the boat is ready to continue digging about noon. Cleaning up was formerly done once in every six days but the two new boats (*No. 7* and *8*) being now completed, it will be done once in eight days.

The clean-up on the *Leggett No. 3* is done differently. The arrangements of the screen and the sluices have been previously explained. The riffles on the upper table are first removed and washed onto the screen, while a full head of water is turned on. The sand and mercury are carried down, the amalgam being retained behind a loose stop. The upper half of the riffles on the second table are taken up and washed and then the lower portion is treated in the same way, one side of the launder being stopped previously, so that everything is carried over one side-sluice. The lower sluices and the 'save-all' are then cleaned up. The whole process only takes about one hour and with the new system of tables it will take even less time. The essential difference from the ordinary method is that the head of water employed while digging is in use throughout the process. The clean-up takes place every ten days and mercury is fed onto the tables and sluices three times between each clean-up. The only boats now using

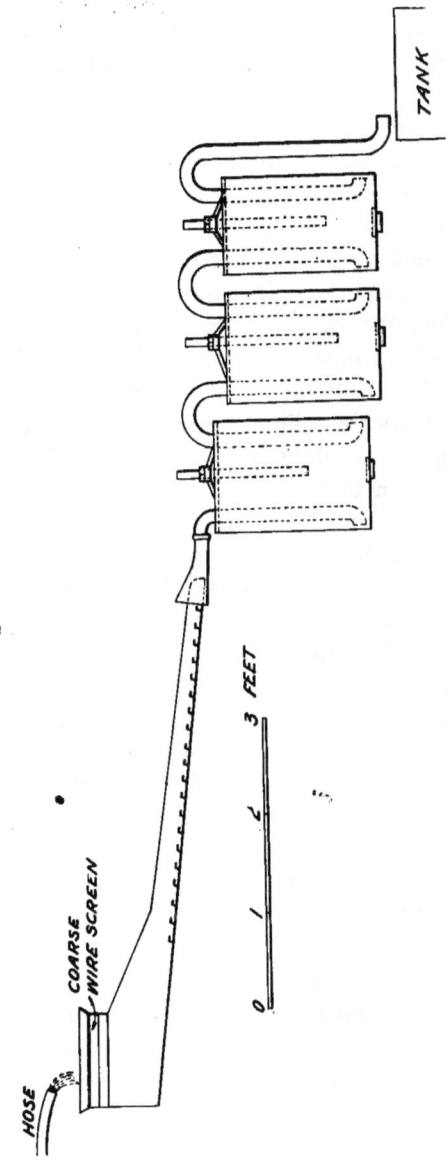

Fig. 78.

the amalgamating machine in cleaning-up are the *Butte* and *El Oro,* and the method as practiced on these boats, though thorough, is slow.

The processes being practically alike, one description will suffice. Each boat has shaking screens with a table underneath with the same slope as the screens and a second underneath the first, sloping back toward the bow and emptying into the launders. Each has a triple divided sluice-way sloping toward the stern, the tail-sluice on the *Butte* being arranged a little differently from the *El Oro* sluice, the riffles of the former being continued a little farther aft. The save-all sluice on the *Butte* discharges into the well of the boat, while the *El Oro* save-all sluice slopes toward the stern and delivers the tailing into a pipe running through the hull. The clean-up, as in all cases, commences at the upper tables first and follows down to the tail.

The slats on each side of the tables are lifted, by removing the wedges, and the sections of riffles are carefully taken out so as not to spill the 'quick' and amalgam that they contain. Then the bottom of the riffle-sections are tapped, scraped, and brushed into a trough, the top sections being done first. The material from this trough and the sand from the tables (from which the riffles have been removed) are conveyed by a couple of Chinamen with pots holding about two quarts, to the miniature 'long tom' (See Fig. 78). In the top of this, as will be seen, is a hopper with a coarse wire screen that may be lifted out and on which the material from the pots is hand-sifted. A hose from the pump is turned into the hopper. Below is a small set of riffles about 15 in. long and 12 in. wide, the overflow from which goes into a 'Lasswell's Fine Gold Amalgamator'. This consists of three iron pots 1 ft. 3 in. diam. and 1 ft. 8 in. deep arranged at intervals of $1\frac{1}{2}$ in. apart, the first pot set on a level higher than the other two. The three are connected by syphon tubes bent into a short elbow at the bottom to give rotary movement to the water and sand, allowing the heavy material to settle. The tops are firmly closed, an air-pipe being inserted in each. A vent in the bottom is closed with a nut. The water, sand, quicksilver, and some amalgam passing over the little tom sluice, go through the series of three pots, the mercury and remaining amalgam being deposited in the first. The material is carried into a tank, which overflows into a launder discharging

THE METALLURGY OF DREDGING.   135

overboard. The 'quick' is drawn off at the bottoms and, later, the tops are taken off and the amalgam is removed. To show the efficiency of the process, the contents of the vat, the result from a month's run of some 40,000 cu. yd., was put through the process a second time and only a little over a desert-spoonful of quicksilver was secured, and this contained very little gold. The material in this vat is replaced near the head of the tables when the riffles are again in place. On the *Butte* the tail-sluices are fitted with one or two sections of cocoa matting, which are rinsed

Fig. 79.   Jets for Shaking Screen, on the Garden Ranch Dredge.

thoroughly in water and replaced, the material deposited in the water being treated with the rest of the clean-up sand. About 50% of the amalgam recovered is found on the riffles in the small clean-up sluice or on the wire screen in the hopper. The contents of each table and the save-all sluice are taken separately and the percentage separately calculated. The results show about 80 to 85% of the recovery from the upper table, 12 to 15% from the lower sluice-tables, and 2 to 3% from the 'save-all'. The last product varies greatly in proportion. On these two boats the clean-up takes place only once per month. From 300 to 400 lb. mercury is put on the table and sluices after each clean-up, and another like quantity during the month. The resultant material from the clean-up consists of amalgam, quicksilver, sand, and small gravel. The cleanest of the amalgam is skimmed, strained through stout cotton cloth, and the quicksilver is used again on the tables, while the amalgam is placed in a metal funnel with a very small hole at the lower end, to drain off the mercury. The sandy portion of the clean-up is panned first over one tub to free it from the coarsest pebbles and practically worthless material, and then over a second. The amalgam and quicksilver are poured off and treated as before, the sand being kept separate.

There are now two products for the melting room: 1. Amalgam with a small excess of mercury; 2. Black sand containing lead, nails, etc. In most of the older districts, particularly those that have yielded large returns in the old gravel-washing days, the material now being dredged contains a large percentage of metal in the form of Chinese and other coins, small ornaments, buttons, nails, and notably lead in the form of shot and bullets, etc. Some of this material is contained in the second product. Much, however, must pass entirely over the tables, acting as a sluice-robber; thus, nearly every dredge has its proportion of lead bullion to be treated after the clean-up, every few months, for this material all collects some amalgam. With the black sand too, are collected traces of platinum, osmium, iridium, and possibly other rare metals. These do not amalgamate in the ordinary way, and it is quite possible that a comparatively large amount passes over the sluices and is lost.

But to return to the melting room: The amalgam, forming the first product, after careful weighing of each portion separately, is placed in an iron vessel and retorted in the furnace. The crude gold from the retort is in its original 'plate' or 'scale' form, though still containing enough mercury to make it adhere in lumps. It is placed with a flux of sodium bicarbonate and borax in a graphite crucible (previously carefully annealed) and smelted in the usual manner. It is then poured into the iron mold, coated inside with the smoke of burning rosin to give the bar a good surface.

The mercury from the dredge is cleaned at intervals, particularly when the boat is idle, by retorting, and leaves a white dross-like looking residue, which carries $16 per ounce in gold;

this is very brittle, possibly by reason of the presence of antimony. In a recent retorting of 963 lb. of mercury, 55.5 oz. of this alloy was produced. It is sometimes melted with the bar and imparts a whitish color to the gold. The second product is screened, to separate the shot and nails, etc. and to eliminate the magnetic sand, the former material being placed in a clean-up barrel with water and iron rollers, and ·rotated. After this cleaning, the nails and iron are removed with a magnet and thrown away, while the residue, including the lead, is retorted to drive off the mercury. The lead matter that is produced is then smelted with flux and the resultant base bullion is worth $1.25 to $1.50 per oz. This is shipped to the Selby smelter with the gold. The black sand that is saved varies in value; that from the *El Oro* and the *Butte* for some six months averaged $140 per ton. Great quantities, of course, go over the tables; no special effort, in fact being made to save it. At Folsom the system is similar to that described on the *Yuba No. 4*. A vat and tom sluice are used and the sandy material from the tom is put through the vat a second time.

With regard to the value of the ground in the different districts, this varies greatly in the same locality, and most of the drilling tests disclose the fact that the highest returns are obtained from channels and patches (probably old river bars), but in dredging a tract the results of the drilling must be averaged to obtain the value of the whole area, which is, as a rule, all moved. For instance, if a rectangular area of 100 acres is intersected by two or three irregular and winding channels with 'islands' between and the values are, on the average, 25c. per yd. in the channels and 8c. per yd. elsewhere over the tract, and the cost of dredging is 6c. per yd., then, though to dredge only the low-grade ground outside the channels would probably not yield a profitable rate of interest, it might be found advisable to dredge the whole tract systematically rather than follow the irregular channels and rich patches even, if when lumped together they formed a fairly large proportion of the whole area. Such questions as these, however, occur constantly to the dredging engineer and must be treated separately and weighed strictly from a business point of view. The vital and only question to be decided is, whether at the completion of dredging a tract of auriferous alluvial land, the operator will have made the maximum amount of profit possible, giving the important factors of time and interest due consideration. As regards the prospect values of dredging ground and the recovery percentages therefrom, in these three districts, reference to the chapter on 'Prospecting' will provide actual cases.

VI. COSTS.

No general table of the cost of a completed dredge can be given on the basis of capacity, as so much is governed by other conditions such as special design, locality, repair facilities, builders, etc. The actual cost of a complete dredge today ranges from $40,000 to $150,000. Even the range in price of boats of the same nominal bucket-capacity varies greatly.

The three prime factors to be kept in mind are:
1. To obtain the greatest possible output in a given time.
2. To maintain the highest percentage of working time.
3. To secure the highest possible extraction.

Fig. 80. 7½-Cubic Feet Bucket on Boston No. 4. Front View.

Boats of 3 cu. ft. bucket-capacity dig on an average of 40,000 cu. yd. per month, and those of larger capacity range over 200,000 cu. yd. per month, as in the case of the *No. 4* boat of the Folsom Development Co., which is fitted with buckets of 13 cu. ft. capacity. This is the largest dredge in the world. The fixed costs of dredging form a large percentage of the total operating cost, and do not vary as much as might be expected with the capacity of the machine. Therefore, it will be obvious how advantageous it is to obtain a large relative output; in fact, production is limited mainly by the mechanical difficulties, and

Fig. 81. Same as Fig. 80, but Rear View.

the increasing percentage of gold lost in boats digging more than a certain yardage. It has been authoritatively stated that with the present type of dredge it is impossible to design a screening and disintegrating arrangement capable of properly treating and preparing for the gold-saving tables over a certain fixed quantity per day, irrespective of the digging capacity of the boat, and that after this ratio between digging and screening capacity has been exceeded, there is an increasing loss in gold per yard dredged. To take an arbitrary case for illustration, let us suppose that in a boat of the most modern design, handling 100,000 cu. yd. per month, the maximum gold-saving efficiency has been reached; then, if the digging capacity be increased to, say, 150,000 cu. yd. per month, the percentage of gold saved per yard will be decreased, though not necessarily in the same ratio as the increase in digging capacity. On this last point hinges the practice on the larger boats at Folsom. Though it is admitted that a certain amount of gold is lost by increasing the output, it is contended that more net profit is made. Another arbitrary example will again illustrate: A dredge is operating 100 acres of gravel property 30 ft. deep and containing 20c. per yd. (prospect value) of which 15c. per yd. is saved at a total operating cost of 7c. per yd., when digging 100,000 yd. per month. The monthly operating cost would then be $7,000. The capacity may be increased to 200,000 cu. yd. per month, with an equivalent loss of, say, 2c. per yd. (probably high) and a decreased working cost of 3c. per yd. To sum up results: In the first case the dredging of the 100 acres, or 4,840,000 cu. yd. is completed in $\frac{4,840,000}{100,000} = 48.40$ months which is 4 years and 12 days, and the gross extraction is $\frac{4,840,000 \times 15}{100} = \$726,000$ at a cost of $\frac{4,840,000 \times 7}{100} = \$338,800$, leaving a net profit of $393,000. In the second instance, where the capacity of the dredge is doubled, the work has consumed two years and six days and $\frac{4,840,000 \times 13}{100} = \$629,200$ has been extracted at a cost of $\frac{4,840,000 \times 4}{100} = \$193,600$, which leaves a net profit of $435,600; a gross saving over the former method of $42,600 in cash and over two years in time, which, if the cost of the dredge was, say, $250,000 would mean an added saving in interest of about $15,000 per year on the investment or $30,000 for the two years, that is, a total profit of about $72,600 over the first case.

The second factor that adds to the cost is loss of digging time, as the fixed charges go on and no gold is being produced to counterbalance them, and, of course, there is the loss in profit and interest. The design and construction of the parts, the character of the ground being dug, and proper facilities for repair, are the chief elements governing this loss, though certain necessary operating conditions and the clean-up also account for lost time.

These last items, are said to be hard to obviate, though in one case mentioned later, it is expected that the time of clean-up will be reduced from the usual period of from two hours or half a day down to as little as half an hour! The following examples from Oroville will give an idea of how time is lost.

1. Lost time on the *Butte* for 32 months was 18.5% of possible dredging, time made up as follows:

Cause	Percentage of time lost
Stepping ahead	9.10
Ladder and bucket-line	21.36
Stacker	6.70
Winches	2.25
Screen	4.40
Water-pump	1.10
Sand-pump	3.81
Lines	2.57
Power	15.70
Other causes (high water and holidays)	27.41
Clean-up	5.60

The sand-pump ran 12.3% of the time.

2. Lost time on *El Oro* for 20 months was 33.41% of possible dredging time; the sand-pump ran for 12.15%.

Cause	Percentage of time lost
Stepping ahead	6.29
Ladder and bucket-line	44.34
Stacker	5.34
Winches	2.34
Screen	12.22
Water-pump	3.90
Sand-pump	1.29
Lines	2.54
Power	3.33
Other causes	15.38
Clean-up	3.30

For the calendar years 1903 and 1904 the lost time on the *Exploration No. 1* was 30.6% and 32.8% respectively of the possible time. The details are most interesting and are as follows:

The best work done by the *Lava Beds No. 3* boat was during the 19th month of operation when an average working time of 22 hr. per day was attained. The 2 hr. per day of lost time included all stoppage of any kind; of course, this is a record and no such average is kept up. On some of the Folsom boats, however, it is claimed that the average working time over a long period amounts to over 20 hours per day.

From the following tables, which refer to six boats of the Feather River Exploration Consolidated, the lost time may be calculated:

Table of Lost Time on Exploration No. 1.
For Year Ending January 1, 1904.

Month	Ladder and Bucket-Line %	Stacker %	Winches %	Screen %	Pumps %	Power %	Lines %	Clean-up %	Other Causes %	Total lost time %	Total dredging time %	Average cu. yd. dredged per day of running time.	Average cu. yd. dredged per day of 24 hr.
January ...	7.1	15.6	5.8	1.0	3.1	5.1	21.6	8.4	32.3	22.3	77.7	1,257	1,000
February...	7.8	0.0	5.0	2.2	9.1	6.5	13.2	0.0	56.2	46.4	53.6	2,007	1,076
March.....	18.4	1.6	1.6	17.6	23.3	14.4	4.9	1.7	16.4	34.9	65.1	1,514	986
April	58.4	7.8	0.7	2.0	0.0	8.0	12.7	2.7	7.7	21.9	78.1	2,100	1,637
May	76.5	1.0	0.0	0.5	1.5	4.1	3.0	6.8	6.6	39.0	61.0	1,900	1,160
June	73.9	4.4	1.0	6.2	0.0	0.7	2.6	7.1	4.2	39.9	60.1	1,890	1,135
July	23.4	5.0	6.2	6.2	1.0	33.0	3.0	6.5	15.7	37.0	63.0	2,114	1,332
August	21.1	9.4	24.3	9.0	2.3	4.4	12.8	11.7	5.0	24.2	75.8	1,958	1,484
September.	20.4	19.7	13.6	1.1	1.0	7.7	11.0	14.6	10.9	24.0	76.0	2,053	1,560
October....	22.3	9.1	9.0	2.0	13.7	7.0	9.0	16.3	11.6	23.0	77.0	2,135	1,644
November .	19.4	18.1	10.0	4.2	2.3	1.5	12.8	11.5	20.2	24.3	75.7	1,752	1,314
December..	20.4	19.0	1.7	1.8	7.8	2.2	12.4	5.2	29.6	30.7	69.9	2,017	1,412
Totals and Averages ..	30.7	9.4	6.6	4.5	5.4	7.9	9.9	7.7	17.9	30.6	69.4	1,875	1,300

Table of Lost Time on Exploration No. 1.
For Year Ending January 1, 1905.

Months	Ladder %	Bucket Line %	Stacker %	Winches %	Screen %	Pumps %	Power %	Lines %	Clean-Up %	Other Causes %	Total Lost Time %	Dredging Time %	Average cu. yd. dredged per day of running time	Actual av. cu. yd. dredged per day of 24 hr.
Jan.	23.9	12.3	9.1	1.6	25.6	4.8	0.4	6.5	2.6	13.2	59.2	40.8	1,365	657
Feb.	6.8	6.8	9.9	4.2	14.9	9.4	23.3	8.2	3.2	13.3	42.8	57.2	1,928	1,103
Mar.	7.9	33.7	5.9	8.1	3.1	—	17.5	9.9	5.4	8.5	32.5	67.5	1,967	1,328
April	6.8	13.2	6.3	—	5.6	0.9	10.5	14.3	5.8	36.6	24.1	75.9	2,229	1,692
May	1.8	21.0	12.9	8.9	12.7	0.7	1.5	11.8	3.8	24.9	22.9	77.1	1,996	1,539
June	0.9	20.6	21.5	1.9	1.4	0.3	14.0	14.3	5.0	20.1	24.8	75.2	1,806	1,358
July	2.5	12.3	26.2	2.1	6.9	0.6	2.6	12.3	4.9	29.6	24.0	76.0	1,883	1,431
Aug.......	0.7	20.1	12.7	0.1	41.4	5.0	9.1	3.6	2.5	4.8	44.3	55.7	2,277	1,268
Sept.	12.4	30.4	6.0	1.7	0.6	2.5	9.1	6.9	2.8	27.6	28.6	71.4	1,942	1,387
Oct.	37.2	12.2	10.2	1.9	0.3	6.8	0.6	7.8	1.5	21.5	25.5	74.5	2,098	1,563
Nov........	25.6	48.9	2.5	0.8	0.3	0.3	4.2	6.6	1.2	9.6	34.5	65.5	2,050	1,343
Dec........	1.8	14.4	5.3	11.2	6.2	0.1	1.4	6.4	2.8	50.4	30.9	69.1	2,166	1,497
Totals and Averages ..	10.7	20.5	10.7	3.5	9.9	2.6	7.9	9.0	3.5	21.7	32.8	67.2	2,005	1,347

DREDGING FOR GOLD IN CALIFORNIA.

Dredge No.	January, 1906. Period.	Actual dredging time Hours.	February, 1906. Month ending.	Actual dredging time Hours.
1	16 days ending January 31	312	February 28	550
2	" " " " "	267	" "	542
3	" " " " "	237	" "	596
4	" " " " "	178	" "	554
5	" " " " "	211	" "	510
6	" " " " "	512	" "	566
Total		1,717		3,318

Dredge No.	March, 1906. Month ending.	Actual dredging time Hours.	April, 1906. Month ending.	Actual dredging time Hours.	May, 1906. Month ending.	Actual dredging time Hours.
1	March 31	517:00	April 30	598:10	May 31	559:45
2	" "	605:40	" "	629:25	" "	490:05
3	" "	589:35	" "	655:05	" "	672:25
4	" "	685:30	" "	310:05	" "	659:20
5	" "	562:05	" "	590:20	" "	561:00
6	" "	613:40	" "	582:50	" "	634.40
Total		3,573:30		3,365:55		3,577:15

The following cases of lost time for two of the boats belonging to the Yuba Consolidated Gold Fields are interesting in comparision:

1. Lost time on *Yuba dredge No. 1* for the 12 months ending January 1, 1906, was 32.8% of possible dredging. time made up as follows:

Cause	Percentage of time lost.
Stepping ahead	5.4
Ladder and bucket line	32.7
Spuds	2.1
Conveyor	6.6
Hopper	2.1
Screen	7.4
Water-pipes and pumps	8.3
Service cables	2.9
Power	4.7
Electrical repairs	1.7
Oiling	2.9
Clean-up	8.5
Miscellaneous repairs (inc. holidays)	14.7

2. Lost time on *Yuba dredge No. 2* for the 12 months ending January 1, 1906, was 30.4% of possible dredging, time made up as follows:

Cause	Percentage of time lost.
Stepping ahead	6.1
Ladder and bucket line	18.8
Spuds	3.2
Conveyor	24.5
Hopper	2.1
Screen	4.5
Water-pipes and pumps	2.1
Service cables	4.0
Power	6.4
Electrical repairs	1.0
Oiling	3.8
Clean-up	12.2
Winches	1.8
Miscellaneous repairs (inc. holidays)	9.5

The bucket-line in every instance, as will be observed by an analysis of the above authentic cases, is the chief time-loser and is the part most often requiring repair. The buckets vary in shape, but the material and strength (other than that due to design) are much the same in every case. The wear on the lips, bottoms, and pins (the chief wearing parts of the bucket), varies so much with the nature of the ground, etc., that citation of particular cases would not be of much use.

The lips, as has been stated, are either of manganese steel or nickel steel. The hoods on the older boats are of forged-sheet steel and often in three sections, riveted together, but practically all hoods, except possibly those of the Risdon-type buckets, are now made of cast steel and thicker than formerly. On *El Oro*, the first lot of buckets supplied, although they had cast-steel hoods and manganese steel lips, were not of proper design nor was the material of good quality, and they only lasted six months. The next lot were of similar material, but evidently of better quality, as they lasted 18 months in the same character of ground. It was found that as the lips wore out from the cutting edge back to the lower end, the angle was altered and the cutting power impaired, so that on the next lot of buckets the lip was flared out at the top and after wearing they had about the original angle in the first case. The lip was originally 9 in. long and $1\frac{1}{4}$ in. thick,

146   DREDGING FOR GOLD IN CALIFORNIA.

but now is 9 in. long and the upper half is 2 in. thick, while the portion that is riveted onto the hood is only 1 in. thick. The present hoods are ¾ in. thick.

In some soft ground near the south end of the Oroville district the hoods and lips in a Bucyrus bucket-line lasted two and one-

Fig. 82. Upper Tumbler of El Oro No. 1, stripped of Wearing Plates.

half years, while an exact counterpart of the above line on the *Lava Beds No. 2*, digging in hard ground, was sent to the scrap heap in nine months.

A Bucyrus dredge at Oroville found it necessary, through wear, to replace all the buckets in an average of 12 months. Another boat in the same district, with precisely the same buckets, has been running already over a year and the buckets are estimated to be good for another year at least. The difference in this case is due entirely to the nature of the ground to be dredged. In the first case the ground was hard gravel strongly cemented together, and in the second, the material was chiefly sand which 'ran' or caved to the buckets, making the work chiefly a lifting proposition.

Fig. 83. Lower Tumbler of El Oro No. 1, showing Wearing of Cushion Plates.

It was thought by the designers of one boat that if the lip was made higher in the centre, it would last longer, the greatest wear being thought to be at that point, but in practice it did not seem to be of much benefit. The usual manganese steel in the lips is made under the Hadfield patent, but the Bucyrus Co. is now supplying a manganese steel lip as well. The Marion Steam Shovel Co. is also supplying a lip of what is termed 'hard tough' steel, which they guarantee to last as long as the other. This is now being tried on some of the new Yuba boats on the same line with bucket lips of the Hadfield steel. Fig. 83 shows buckets with Hadfield manganese steel that have been digging for 22 months, and are still in use on the *Yuba No. 2*.

On the *Ashburton* at Folsom the bucket lips are of nickel steel, 8 in. long by 1¼ in. thick, and forged at the company's shops.

Fig. 84. Bucket Line of the Ashburton Dredge. Lips of Nickel Steel.

On the latest and largest Risdon boat, the *Baggette*, just completed at Oroville, the bucket lips are of both nickel-tool steel and manganese steel, and are 2 in. thick on the digging face, 1 in. at the back and 11 in. deep; the hoods and bottoms are of flange steel, the former ⅝ in. thick, and the pins and bushings are of manganese steel. The Risdon bucket still retains its deep and somewhat narrow shape, while the modern tendency seems to make for wide and shallow buckets of large capacity that will dump cleanly over the upper tumbler.

Pins are a source of expense, and have undergone many modifications and changes. In one of the Lava Beds Co.'s boats pins of so-called projectile steel, made by the Risdon Company,

Fig. 85. Idler Drum. Diameter 9 feet 6½ inches.

lasted for twelve months in hard cemented gravel, while Fig. 85 shows a pin and bucket-bottom from the *Yuba No. 1,* which were discarded after eighteen months constant service. In this case the dotted lines show the original outline of pin and bottom at wearing parts; the eye of the bucket-bottom, as may be noticed, is worn down to almost the thinness of a knife-edge.

Fig. 86. Showing Wear of Bucket Bottom and Pin after 18 Months Service on Yuba No. 1. Dotted Lines show Original Outline.

Instructive records have been kept of the life of various pins on the *Butte* at Oroville, as follows:

Material	Wear Months
(Original pins) Soft machined steel	6 (extreme).
Rolled steel with welded heads	(Said to be very poor.)
Marine steel from the Boston Machine Shops	12 to 14.
Marine steel from San Francisco	6 to 8.
El Oro shops, tool steel	5.
Cambria Steel Works	10.

At another locality, common shafting was found to wear from four to six months, and Harveyized armor-plate lasted twelve months and over, but could only be machined with difficulty, as it was too hard. The pins on the Lava Beds boats are now made in three different sizes, 3⅞, 4¼, and 4⅜ in. diam., and are made of so-called marine steel at the Boston Shops, the larger sizes being used as the bushings wear. On the *Exploration No. 2*, also a 5-cu. ft. bucket boat, the pins of the same material and make, are 3 13/16 in. diam. The manganese steel pins on the *Baggette* are 3 in. thick, and solid. On the *Exploration No. 7*, a Risdon boat, which has changed its bucket-line for a Bucyrus close-connected line, the pins are 2 63/64 in. diam., and are of machine-steel. A forged-steel pin of 3⅞ in. diam. was used on *El Oro*; and on this

Fig. 87. Bucket Line, showing Wear of Lips after 22 Months Continuous Service.

and several other boats a solid manganese pin of about the same diameter has been replaced by a hollow manganese steel pin of 4½ in. diam., with a hole of 1⅛ in. through it. This last pin has been in use for a year with the 5-cu. ft. buckets on *El Oro,* and seems to give perfect satisfaction. Some Harveyized pins on the same boat have been on for a year and a half, and are still in use. At Folsom the pins do not differ materially in size or material from boats of equal capacity, while on the *No. 4* boat of the Folsom Development Co. the pins are of 3% nickel-steel.

At one of the old Risdon-type boats at Oroville, I observed an interesting operation: The bucket line (3 cu. ft. capacity) was lying on the bank, and my attention was drawn to the scene by a number of small explosions. On arriving at the spot I found that nearly all the pins in the line, which were only slightly over two inches in diameter, were broken, and the method of abstracting the ends was unique. Two pieces of flat-iron bar were driven in at the link between the broken ends of the pin, and a small charge of dynamite being exploded between the bars, the force drove the pins partly out at the ends of the eye. In this same bucket-line a majority of the lips were cracked—in some cases at several points and directly across, only being held together by the rivets through the hood.

Pins vary considerably, in first cost, according to their composition. The hollow 4½ manganese steel pins on *El Oro* cost $25 each; the Harveyized steel cost $45, and the ordinary high carbon forged steel cost $16, so that the price must be considered as well as the life.

The bushings used are made semicircular because the wear is practically all on the rear end of the eye. They are shown in Fig. 86, and are almost universally of manganese steel and about ⅝ in. thick. An interesting innovation in the method of making these was shown to me by Mr. Krug in designing a new bucket-bottom, the object of which was to obviate change in pitch* distance and thereby lengthen the wear of the line. The bushing is made slightly thinner at the centre, and thus on wearing at the edges to the same thickness, an ordinary bushing may be put in.

Solid cast iron or steel is used by some dredgemen for ladder-rollers, and others use a special quality called 'semi-steel'; they may also be made hollow. On *El Oro* the first lot of rollers were of ordinary cast iron, and lasted one year. Rollers of 'semi-steel' were then put on and have already lasted fifteen months, and appear to be hardly worn at all. They are placed 7 ft. apart. Solid rollers are sometimes as small as eight inches in diameter, while hollow rollers run up to 14 in. The best results, however, appear to be given with manganese steel rollers, which preserve their surface and do not wear the bucket-bottoms unduly. At times the rollers stick, and it is remarkable in such a case how quickly they will wear down flat.

It is difficult to obtain the correct relation between the ma-

---

*The 'pitch' of a bucket-line is the distance between the centre or one rear 'eye' and that of the rear 'eye' in the bucket-bottom immediately behind it when coupled. The constant wear on the pins and bushings changes this distance, and allows more 'play' and consequently more wear in the whole line.

Fig. 88. Wear Plates for Lower Tumbler.

terial used in the bucket-bottoms and that in the wearing-plates and shoes on the tumblers. At present the former is generally made of high carbon steel, and the latter of manganese steel or nickel-steel, and one bucket-line will wear out several sets of 'cushion' or wearing-plates on the tumbler. On the Yuba boats

Fig. 89. Wear Plates with Lugs for Upper Tumbler.

nickel-steel shoes and plates are used on the upper tumblers, and last for six months, while the lower tumbler wearing-plates are of manganese steel, and though they last over a year the tendency is to break rather than wear. It might be at first supposed that the proper relation would be to have wearing-plates last the same period as the bucket-bottoms, but this is not so. A set of wearing-plates will cost, say, from $600 to $700, and a bucket-line between $10,000 and $30,000, so that the object is thus at all hazards to lengthen the life of the bucket-line. It is not a difficult job to replace wearing-plates, and does not occasion an undue loss of time. Moreover, there is a wearing action from slippage of bucket-bottom on tumbler-face (caused by grit) such that if an even relation between bucket-bottoms and wearing-plates existed, the life of the former would be materially shortened at a greatly increased annual expenditure.

On screens the ordinary repairs may be done during clean-up time, and this should also be utilized in making all other repairs that are possible.

The relative value of the belt-conveyor as against the pan-stacker for removing the coarse oversize from the screens to the stack-pile, has been already touched upon. A few interesting records of the relative wear of belts are available.

On the two boats at Oroville several varieties of belt were used, the life of which are given herewith :*

Name of Belt	Time of Wear Days
Robins belt (Robins Conveying Belt Co.)	226
Bowers Rubber Belt (Bower Rubbers Co.)	219
†West Coast rubber belt	320
Gutta Percha Rubber & Mfg. Co.	174
Link Belt (Manhattan Rubber Co.)	112
Gardner Belt (Boston Woven Hose & Rubber Co.)	252
" "	287
" "	330

*Information from W. S. Noyes.
†Using two pad belts.

On *El Oro* a Peerless belt was in use, with pad belts, for 18 months:

	Days.
Ran without a pad	150
With 6-ply ordinary 16 in. drive belt	17
With special pad ⅝ in. thick with ⅛ in. rubber on each side, 16 in. wide	137

	Days.
*16 in. Gandy belt	26
†Balata belt, 16 in.	150

*Canvas (soaked in oil), contracted in width and stretched in length.
†An English belt—worn so thin finally that lacings could not be kept in.

The cost of labor and material for the pan-stacker on the *Exploration No. 1* at Oroville for the twelve months ending January 1, 1905, was $2019, while for the previous year it was only $1241. Mr. James H. Leggett states that on his 5 cu. ft. boat the annual cost of maintenance of the pan-conveyor is between $300 and $500. On the Syndicate dredge at Folsom the labor of repairing the pan-stacker cost an average of $10 per day.

A new application to dredges in the way of conveyor-belts is the Ridgway, which is being tested on *Yuba No. 4*, and consists of an inside canvas belt with curved wooden blocks or carriers, in which runs the upper belt that carries the material. A double driving gear is used with motor at the outboard end of the stacker-ladder. On the shafting of the drums of each belt is a different sized gear-wheel run by the same pinion and arranged so that they run at the same speed, and slipping (of one belt on the other) is obviated. A new belt is being introduced on the dredges, which it is claimed will give superior wear. The belt is driven full of staples, which are clinched on the carrying side of the belt, and arranged in rows that break joint. It is said that this will give extra strength as well as durability.

Cases of broken spuds are comparatively few, but it is doubtful whether the practice in use at the Champlin dredge at Footes creek in Oregon might not be of benefit in the Sacramento valley. Here it was found cheaper to make both spuds of wood; when one breaks it is simply exchanged for another, and the great loss of time entailed in repairing a broken steel spud is obviated. Of course, suitable sticks must be kept on hand and ready for immediate use. In some hard ground, however, wooden spuds would be entirely out of the question.

Modern dredges are equipped with steel traveling cranes at several positions on the boat, and these are almost indispensable in the great saving of time over the use of blocks or in rigging a temporary derrick every time a repair or replacement of any consequence is to be made.

It has been stated by one writer on dredging that records of lost time and costs for the first few years on a dredge should not be taken as accurate. This point has been emphasized unduly by several dredging critics. As a matter of fact, though the average over a long period is a better criterion of the cost of operating, as would be the case in any other industry or business, if breakages are honestly repaired at the proper time and in an up-to-date

Fig. 90. Ladder Hoist on the Baggette.

manner, a record at the end of the second year should be fairly accurate. For instance, nowadays most of the parts subject to constant breakage are generally sectionalized to such an extent that the actual wearing parts may be replaced separately, and the entire machinery and plant of a dredge might at the end of five years be actually better than at any time previous; it might have been necessary to entirely renew every part during the fifth year, some of them, of course, having been renewed several times before. Moreover, the material used, the workmanship, and the design

Fig. 91. Showing Wear on Lips of El Oro Buckets.

would all be subject to improvement during the five years' experience. The only part of the dredge that might have shown wear at that time would be the hull, but these, in modern dredges are staunchly built and in such a manner that they may outlast the area over which they are designed to operate.

An important element in the saving of time, and therefore in cutting down operating costs, is the presence of good railway connections and the nearness of facilities for making repairs. At Oroville there are several shops of different capacity, in some of which practically all the work required on a boat may be done, except the manufacture of castings. These shops are owned by the several companies, some of which do custom work as well. The Boston shops are the largest, and are well fitted for repair and construction work, in fact, for several years they have been constructing the boats of their company, both at Oroville and on the Yuba. The equipment consists of two traveling cranes, an air-compressor, 60 by 60 in. Detrick & Harvey open side planer, 6 by 6 ft. Cincinnati planer, four radial drills of varying sizes, horizontal boring mill, shaper, double traverse head shaper, 8-in. pipe machine, 6-in. pipe machine, lathes, grinders, small tools, etc. The equipment also includes a steam hammer, cut-off shears, oil furnace for heating plates, blacksmith shop with seven fires. Also a very complete line of pneumatic tools, chipping hammers, riveting hammers, drills, etc. A notable feature of the equipment is a compressor and motor mounted on a wagon and fitted with air hose and complete equipment of pneumatic tools, which, in case of trouble, may be quickly hauled to a dredge to facilitate repairs. The shops cover an area 80 by 160 feet.

The larger repairs on the Yuba dredges are done at the Boston shops, and minor work at their own temporary shops at Hammon City. These will soon be equipped with a steam-hammer, hydraulic press, large forges, radial drill, a compressor-plant, and lathes, etc. Arrangements are also being made for the erection of extensive shops at Marysville, which will be so fully equipped that a dredge may be entirely constructed there.

At Folsom, the Folsom Development Co., has the largest and most elaborately equipped machine-shops for this work in the State, outside of San Francisco, and every possible repair (except castings) is expeditiously made here. In their store-rooms, it is aimed to keep an adequate stock of all articles, the ordering of which at a distance would entail delay in shipment. Weekly reports are made to the business office of the stock on hand so that replacements can be quickly made. The necessity for this can be understood when it is stated that it often takes seven and eight months to fill orders for some of the larger castings.

A vivid illustration was presented at Oroville only a month ago, of loss on account of not having parts on hand. A 5-cu. ft. bucket boat broke an upper tumbler shaft, which had to be sent to San Francisco for repair. Over four weeks were consumed before the boat was running again, and during this time a large proportion of the fixed charges went on, as well as interest, and no income was available. Two days at the outside are all that would have been necessary had a spare shaft been kept on hand, and the interest on the money so expended would be small compared with the loss.

The fixed costs of dredging include office management and bullion charges, labor, power, water, taxes, insurance and interest, and probably 50 to 60% of the total operating cost per yard.

The bulletin published by the California State Mining Bureau contains many instances of cost, principally in the Oroville district, and quotes a table contained in a paper read before the California Miners' Association by Mr. L. T. Hohl which I append.

*Costs of dredging at Oroville (in cents).

Item	No. 1	No. 2	No. 3	No. 4	No. 5
Power	1.06	1.20	1.15	1.61	1.77
Repairs	2.86	3.03	3.46	2.97	3.80
Labor	1.64	1.32	1.85	2.33	2.05
General expenses	0.64	0.67	1.23	1.28	0.73
Total	6.20	6.72	7.69	8.19	8.35

Whether these include the expenses of office, managements, taxes, interest, insurance, which in many of the other cases given seem relatively very small, is not stated.

It seems to be conceded, however, by the best practical dredgemen that the average cost per yard at Oroville for a year's work

---
*These referred to operations several years ago.

does not get much lower than 6½ to 7c. Indeed, I am aware of at least one instance during the past year where the cost exceeded 12c. It is probable, too, that the yardage value of much of the ground has been quoted too high. The following costs are for the *Exploration No. 1* for two consecutive years. It must be borne in mind, however, that this boat is somewhat out of date, and not of large capacity.

Item	For year ending Jan. 1, 1904.	For year ending Jan. 1, 1905.
	Cents per cubic yard.	
Labor		
Operative	1.37	1.29
Repairs	0.61	0.94
Superintendent	0.38	0.38
Power		
Dredge	1.08	1.00
Pumps	0.07	0.01
Hardware		
Supplies and tools, etc.	0.32	0.22
Repair parts	2.32	1.37
Freight and express	0.04 ⎫	0.10
Hauling	0.07 ⎭	
Steel cables	0.07	0.07
Lumber	0.03	0.006
Electrical supplies	0.005	0.06
Clearing ground	0.095	0.16
Sundry expenses	0.02	0.007
General plant	0.24	0.26
Bullion expenses	0.04	0.05
Prospecting	0.12	
General expense		
Oroville	0.38 ⎫	0.36
San Francisco	0.17 ⎭	
Taxes warehouse, insurance, legal	0.28	0.77
Total	7.71	7.06

While, for the two years the total costs are close, the monthly total costs per yard varied between 4c. and 14c. This dredge cost $45,000 and dredged in 1903, 474,610 cu. yd., and in 1904, 493,150 cubic yards.

For the past 20 months the *Lava Beds No. 3* of the Oro, Light, Water & Power Co. has dug an average of 86,330 cu. yd. per month at an average total cost of 4.77 cents per yard.

The total costs per cubic yard for the six boats of the Feather River Exploration Consolidated for the five months ending May 31, 1906, were as follows:

Cost in Cents Per Cubic Yard.

Month	Boats					
	No. 1	*No. 2*	*No. 3*	*No. 4*	*No. 5*	*No. 6*
January	3.91	5.16	5.88	10.06	4.15	3.70
February	4.86	5.15	2.41	9.72	5.82	4.55
March	4.44	6.70	3.13	4.35	2.79	4.39
April	4.52	4.60	2.73	15.52	2.82	4.33
May	5.76	9.61	4.71	5.50	5.93	5.13

No. 1 is a Risdon boat of 3½ cu. ft. bucket capacity. *No. 2, 3, 4* and *5* are also of Risdon construction with 5-ft. buckets, and *No. 6* is a Bucyrus with a close-connected 5-cu. ft. bucket line.

Probably at Folsom and on the Yuba the best work is being done in dredging today. Assisted by the examples of others' mistakes, and by the development at Oroville and elsewhere, boats of huge capacity have been constructed. This has been accomplished after careful study, with every possible precaution that the present stage of the practice can suggest, at the company's own shops, and under the direction of their own engineers. The resulting machines actually dig from 100,000 to over 200,000 cu. yd. per month.

At Folsom cheap power comes into play, and the cost being only 65c. per kilowatt hour, a saving is at once effected of over one-half the power cost at Oroville, equal probably from one-third to two-third cents per cubic yard.

By proper organization and office management, thorough facility for repairs, and the following of a well planned system, expense may be reduced. An important element in this reduction is the operation of as many boats as possible under one management. One general, departmental, assay, and other offices, one repair plant and store-house, with a reduced number of heads of departments and employees will cut down the salary, building, and maintenance accounts immensely. Labor, of course, is one of the important fixed charges in dredging. Usually three shifts are employed, and the regular crew consists of a dredgemaster, winchmen, and two oilers. The wages are as follows:

District	Foreman or Dredgemaster	Winchman	Oilers	Other Labor
	Per Month	Per Shift	Per Shift	Per Shift
Oroville.......	$125	$3.00	$2.50	$2.00
Yuba..........	$150	$4.00	$3.00	$2.50
Folsom........	$150	$3.50	$2.75	$2.50

Special labor is sometimes required at the clean-up, on repair work, and always in clearing the ground, blasting, prospecting, etc., and all this must be added to the annual yardage cost in computing the profit.

The cost of power is more or less fixed, but it may be reduced by care and the use only of the most improved forms of motors, cables, winches, and driving gear, etc.

The cost of water is not such an important item and depends entirely on the local conditions. Ingenuity in the use of stream and river water may, however, save the purchase of many inches of irrigation water. On the Yuba the only difficulty in this connection is in pumping water out of the ponds, as the seepage gives the required supply.

By referring to the tables of cost, the relative expense of power may be deducted. The cost at Oroville is $1\frac{1}{2}$c. per kilowatt hour, the horse-power used varying from 60 to 275 per boat. The nominal horse-power may be obtained from the detailed description of boats given under Chapter VIII.

[Further details of cost are given in the Appendix.]

Fig. 92. Belt-Conveyor on Stacker-Ladder of Folsom No. 3, fitted with Sets of Four Idlers.

VII. THE HORTICULTURAL QUESTION.

Much has been said and written since dredging commenced in earnest in the Sacramento valley on the question of the destruction of the land by rendering it unfit for future agricultural purposes, chiefly fruit culture. A large portion of this matter, particularly that which has appeared in certain popular magazines, is, to say the least, misleading. Efforts have actually been made to restrain companies from carrying on their operations. No apology therefore is needed in referring to this subject. Care has been taken to obtain the facts from persons interested both in dredging and in the culture of fruit, and the following brief remarks sum up the results of such investigation.

Oroville is the most important district in this connection and here the possible dredging lands, including what have already been worked, will probably cover about 6000 acres. Of this tract less than 400 acres has been planted in fruit and though about 1000 acres of the remaining portion, at an outside estimate, might, with irrigation, be used, it is ground of a poorer quality, requiring much annual expense in water and cultivation, and it is extremely doubtful, judging by the prevailing conditions, whether it would ever be used for this purpose. Of the 400 acres mentioned, 150 acres are included in the Leggett vinery, which for several years has been infested with the practically fatal phylloxera and probably $300 per acre would cover the amount on which it would pay, say 6% per annum, for horticultural purposes. Similarly, the Wilcox peach orchard, covering 50 acres, produced an average of one crop in three years, and might provide an income of 6% on $100 per acre. The Gray peach orchard of 20 acres would pay a fair rate of interest on $200 an acre. The Gardella tract is a fine piece of land of 40 acres and peculiarly adaptable for the production of vegetables as well as fruit, and for market gardening generally, and would possibly pay 6% on $1000 per acre. The remaining portion cannot be valued at over $50 per acre, and the whole 6000 acres would not represent at the outside an invested capital of over $250,000. Moreover it is a well-known fact that most of the ranches in this district were mortgaged.

If we say that the district will yield on the average 10c. per yd.

net profit and averages eight yards in depth, both fairly conservative estimates, then over twenty-three million dollars net profit will be produced and if invested in the State at 6%, will produce an annual income of over $1,380,000 instead of the paltry revenue of $15,000 per year, which the land produced before. Besides this for a long period a large population will be employed in the district and at the end of the period of operation another twenty million dollars should have been expended in labor and machinery.

On the Yuba the question of interference with arable land does not arise.

At Folsom most of the bench land has been worked over before by gravel miners, who left a large proportion of it in such a condition that it is unfit for cultivation. Of the 6000 acres available for dredging, probably not over 5% has been planted and little more is available for planting, the chief product being grapes. This land is owned by the Natoma Vineyard & Winery Co., and the population it actually supports is small. A comparison similar to that at Oroville might be made out regarding this district, which would probably show even less favorable results for those busybodies and other uninterested parties who have attempted by unwarranted agitation to raise a storm in a tea cup. If the question be really pushed to an issue, an excellent solution would be to invest enough of the proceeds of the dredging operations in the filling of the now worthless tule lands (of vast extent in this valley), so that an acreage equal to that destroyed might be put into cultivation. This, however, would probably be an inconsiderable sum, particularly as the sale of such land would, of course, revert to the dredging companies.

After dredging the land is covered with stack piles containing boulders and pebbles, the fine material being hidden 6 to 15 ft. below. In regard to the reclamation of land in this condition, certain experiments attempted by Mr. J. H. Leggett at Oroville, produced interesting results. Eucalyptus of both the 'blue' and the 'red gum' variety and some fig trees were planted on the stack piles, and alfalfa grass is also growing. The trees, when set out, were only about 10 in. high. Fig. 93 shows the results at the end of two years. In the case of the largest tree, 19 ft. of growth was attained and of course this will increase in greater proportion with time. Some of these trees were planted directly on top of

166   DREDGING FOR GOLD IN CALIFORNIA.

the highest ridges, 15 ft. above the original surface and Mr. Leggett states that one cannot dig a hole at any of these highest points more than 2 ft. without getting moisture, even in the driest time in summer. The volcanic ash that forms the bottom of the

Fig. 93. Eucalyptus Two Years After Being Planted on a Stack Pile. 19 Feet High.

digging and the material of certain clay bands is often carried over the stacker in lumps and disintegrates on being exposed to the air so as to form the soil necessary to growth. The alfalfa roots were accidently thrown up by the dredge and the grass

THE HORTICULTURAL QUESTION. 167

Fig. 94. Eucalyptus Just Planted on Stack Pile. 8 Inches High. See Pipe.

seems to do particularly well. Absolutely no care was given to the trees or grass from the time they were set out.

Nevertheless, with the stack piles in their present condition of irregular surface, it is not considered that a profitable horticultural industry could be built up. It has been suggested that the piles might be leveled with huge scraping rakes run by electric power, and in this connection an instance of the present cost by hand for leveling at Oroville is given. Near the town the original owner of a tract that was dredged agreed to put the ground in a suitable state of cultivation if given to him. He leveled one acre and covered it with one foot of soil and now has a productive garden; the total cost of leveling and covering being $250.

Ground that was dug by the single-lift dredges (with the long sluice on a pontoon to the rear) was left in as good condition as before dredging commenced, that is, for agricultural purposes. It is perfectly flat and the fine tailing is mixed with the pebbles and boulders, and in many cases is altogether on top. Instances of this may be seen where the Continental dredge operated at Oroville, and at Folsom a particularly good example is afforded by the work of the *Ashburton*. This boat originally employed the long-sluice method, but changed to the ordinary screen and tables, and the contrast between the two results in stacking may be observed on the same property, the work having been done by the same dredge.

At Folsom the stack piles are to become a valuable asset in the hands of the enterprising management of the Folsom Development Co., who are installing a rock-crusher plant and the product is to be used in railway and road ballast, and for concrete work.

Fig. 95. The Foreground Shows the Flat Tailing Bed of an Old Dredge, while the Stack Piles in the Background show the Work of the Modern Boat.

VIII. GENERAL.

A few brief descriptions of each district in regard to the number of companies and boats in each, nature and value of ground, etc., are given below.

There are sixteen companies now engaged in dredging at Oroville and thirty-two dredges are operating, as follows:

Name of company	No. of Boats	Builders
American Gold Dredging Co.	2	Bucyrus.
Butte Gold Dredging Co.	1	Bucyrus.
El Oro Dredging Co.	1	Link Belt Machinery Co.
Feather River Exp. Con. Co.	5	Risdon.
" " " " "	1	Bucyrus.
Gold Run Dredging Co.	1	Bucyrus.
Indiana Gold Dredging & Mining Co.	2	Bucyrus.
Leggett Gold Dredging Co.	1	Risdon.
Nevada Gold Dredging Co.	1	Bucyrus.
Ophir Gold Dredging Co.	1	Bucyrus.
Oro Water, Light & Power Co.	1	Risdon.
" " " " "	3	Bucyrus.
Oroville Dredging, Ltd.	4	Risdon.
" " "	3	Bucyrus.
" " "	3	Marion Steam Shovel Co.
Oroville Dredging Co.	1 (dipper)	" "
Pennsylvania Dredging Co.	1	Golden State & Miners' Iron Works.
Pacific Dredging Co.	1	Bucyrus.
Viloro Syndicate	1	Bucyrus.

Formerly this district was one of the most important placer camps in the State, the second discovery of gold having been made a few miles above the present site of the town in 1848 by John Bidwell. The richest portion of the gravel having been worked over by white men, Chinese took their place and swarmed over the ground, working the same claims over again in many instances. In fact it is these very portions, already worked over several times, that now give the highest returns on the dredges. It is generally supposed that the Chinese did not work below the water level but in many cases there is abundant evidence, both in the drilling and

Fig. 96. Remains of Old Workings on the American River at Folsom.

dredging operations, that they did. Old workings are frequently encountered in drilling and timbers from these tunnels and wing-dams, etc. are often brought up by the dredges. The Chinese

Fig. 97. Old Chinese Workings at Folsom, showing Home-Made Water-Wheel for Chinese Pump.

pump was employed and actuated either by a small local water power or by hand. This contrivance consisted of one or two 3-in. belts with blocks of wood attached as elevators, and a home-made water-wheel, with board launder from an irrigation ditch

GENERAL. 173

ɔr stream, completed the plant. Fig. 97 shows a typical Chinese plant as seen today, in ground worked over years ago and probably to be worked again, by the dredges.

At Oroville I saw a link with the past. Within sight and hearing of the creaking and groaning dredges, sheltered from the summer sun by a few boards, sat an old man from the Oddfellows home nearby—a relic of the days of '49. He was working a rocker and an occasional trip with a wheelbarrow to the neighboring bank

Fig. 98. Gravel Bank at Oroville.

would provide him with gravel for most of the day. Though far from communicative I learned that he had, during the fine days of summer in the past three years, driven the little tunnel shown in the accompanying illustration. It was only about 8 or 9 ft. long and during that period he had washed some 13 or 14 cu. yd. of gold-bearing gravel.

The gravel in this district is from 18 to 40 ft. deep and averages about 30 ft.; the bore-hole sections given under 'Prospecting' will give an idea of its composition. The bottom of the dredging

174   DREDGING FOR GOLD IN CALIFORNIA.

ground is generally a volcanic ash, quite soft, but often found to be tough and elastic in dredging. As to gold being contained in appreciable quantities below the present digging level, as has been stated by some writers, the result of the deep bore-hole, already quoted, speaks for itself. It is not known that any gold-bearing stratum of any consequence lies underneath the level to which digging is being carried in this district and, contrary to general sup-

Fig. 99.   An Old Miner and His Rocker.

position, the gold in the gravel is not evenly disseminated from top to bottom over the entire district. It would indeed be an extraordinary case if it were so. (See 'Prospecting.')

It would be a difficult matter, with the present lack of information and the unwillingness of the dredge-owners to make it public, to give correctly the average value of the gravel per yard, either in this or the other districts. It has been variously stated to average from 15 to 25c. per cu. yd., and probably a correct average would fix it much nearer the former figure than the latter, for the whole 6000 acres. This figure depends of course upon whether the "prospect" or "recoverable" value is taken, the latter being, as a rule, considerably lower than the former. In an area of some 560 acres, 175 holes gave an average result of 14c. per cu. yd. Another tract of 200 acres near the town of Oroville prospected an average of 36c. per cu. yd., while an adjoining property of over 100 acres was said to average 28c. The ground near the town of Oroville appears to be fairly rich, while the tracts at the extreme southern end of the district are generally of lower grade, but the character of the ground renders it easier to dredge. Consequently the cost at the southern properties have been as low as 3c. per cubic yard.

The power used is entirely electric and is controlled by the California Gas & Electric Co., which has dams, ditches, and power plants at several points in the higher foothills. A good irrigation system exists and supplies water where necessary.

A brief description of some of the boats, including the newest and oldest of each type, at Oroville, will give an idea of the difference in design.

The housing of the earlier Risdon boats usually had pitch roofs and the trommel and gold-saving tables were not housed in. An example is the *Exploration No. 1*. The hull is 86 ft. long by 30 ft. wide, and 7 ft. deep. The bucket-line was originally open-connected with buckets of 4-cu. ft. capacity. This has been changed to a close-connected line of 78 buckets, each of 3-cu. ft. capacity. The ladder is made up of plate and angle-iron girder sides with bulkhead plates and lattice bars between. The trommel screen is 25 ft. in length and 4 ft. 6 in. diam.; the gold-saving tables, 12 in number, slope from the screen toward each side, emptying into a stream-down sluice. Angle-iron riffles are used. The stacker is composed of a pan-conveyor and the boat works on a head line.

The latest Risdon boat built is the *Baggette*. The hull is 98 ft. long by 34 ft. wide. The ladder is of truss-girder design and the open-connected buckets, of which there are 39, hold 7 cu. ft. each. The upper tumbler on all Risdon boats, is square and the lower tumbler is pentagonal. The middle gauntree is built of iron, but instead of the usual design—battered fore and aft—the back is vertical, thus giving more room aft for the screening apparatus. An improvement has been made also in the driving gear. It has always been held by some that the use of a large gear-wheel on one side with a small pinion below is a source of trouble and weakness, particularly as the bearings of gear-wheel and pinion rest on separate castings. This has been obviated in the *Baggette,* and the pinion is more nearly to one side of the gear-wheel, and in the same casting. A new arrangement was tried for hoisting the ladder. The motor was placed at the head of the forward gauntree and two chains running over sprockets were used, the ends of the chains dropping into the hold. It was found not to be a success and it was replaced with a sheave-wheel arrangement, similar to the ordinary ladder-hoist. The screen is 20 ft. 6 in. long and 6 ft. diam. and both it and the gold-saving appliance have been thoroughly described in Chapter VI. The motors are as follows:

Purpose	Horse-power
Main drive or digging	75
Ladder hoist	35
Pumps	50
Priming pump	5
Screen	20
Stacker	15
Winches	20
Total	220

The boat, like all others of the Risdon type, is operated on a head line. Six men are employed on three shifts, not including the dredgemaster.

The *Boston No. 4,* although chiefly built by the Boston shops at Oroville, is a modified Bucyrus type. An improvement has been made in the design of the forward gauntree frame over the old 'A' style and four parallel upright posts are used, strengthened with a good system of braces and securely tied to the main framing of the boat. The cap is of steel. The gauntree slopes forward at an angle of $5\frac{1}{8}$ in. horizontal to 1 ft. vertical—somewhat steeper than usual. The digging ladder is very heavy and constructed of plate and angle-girder, the sides braced with lattice-bars between. The bucket line consists of 68 close-connected buckets of $7\frac{1}{2}$-cu. ft. capacity each. They dig at the rate of about 20 per min. The upper tumbler is pentagonal and the lower one hexagonal.

The trommel is 6 ft. diam. and 30 ft. long, 28 ft. of which are perforated. The gold-saving arrangement is as follows:

The stacker is of the belt-conveyor pattern and the boat is operated on spuds. The motors are of the following nominal capacity:

Purpose	Make	Horse-Power
Main drive and ladder hoist	General Electric	150
Low pressure pump	" "	25
High " "	" "	50
Priming "	" "	10
Side line winches	" "	25
Screen, spuds, and stacker drive	" "	25
Sand pump	" "	75
	Total,	360

On the Yuba there are two companies: The Yuba Consolidated Gold Fields and the Marysville Gold Dredging Co., the first with eight dredges and the second with two. These companies, however, are practically under the same general management.

The headwaters and tributary streams of the Yuba river have for years been mined extensively by hydraulicking and it is estimated that beside the natural detritus, carried down each year in the floods, there have been 300,000,000 cu. yd. deposited in the lower river-bed alone.

Naturally the old river-bed gravel, which is gold-bearing, has been covered with an overburden consisting chiefly of sand and 'slickens' from the hydraulic workings; this varies from 12 to 30 ft. in depth and though it contains some gold, it is difficult to discharge clean. The total digging depth is from 50 to 70 ft. Below the overburden of tailing is the usual 'wash', similar to that of the Oroville district. Below this, the volcanic ash is encountered and presents the same sticky and elastic mass so difficult to dig that forms the so-called 'false bedrock' in the other districts. Boreholes below this showed some gold down to 115 ft. from the surface.

With the exception of *No. 1* and *No. 2* of the Yuba Consolidated Gold Fields, the boats of both companies are of the same type. *No. 1* and *No. 2* have buckets of 6-cu. ft. capacity, while those of all the others are of 7½ cu. ft. They are all of the close-connected type and all the boats operate on spuds and, except *No. 1* and *No. 2* (which each have two shaking screens, in extension) use revolving trommels. The most marked impression a new observer receives of the Yuba practice is the immense length of stacker necessitated by the great depth of the ground dug. A description of the *No. 7* boat of the Yuba Consolidated Gold Fields will typify the best points in the others. The hull is 115 by 40 ft. The digging-ladder is of the open truss-girder design, 114 ft. long and the bucket-line contains 93 buckets close-connected and of 7½-cu. ft. capacity each. Both the upper and lower tumblers are hexagonal and the upper tumbler-shaft is 17 in. diam. at tumbler and the journals are 13 in. diam. The trommel is 30 ft. long and 6 ft. diam. The gold-saving tables on both this boat and *No. 8* were intended to be on the Holmes system, but unfortunately these were destroyed in the San Francisco fire while they were being manufactured. Accordingly, the old pattern side tables with stream-down box, were put on *No. 7* and Holmes tables, partly wood and partly of iron, were improvised for *No. 8*. It is intended later to change the tables on *No. 7*. The spuds on these boats are 55 ft. long and 24 by 36 in. section; they are constructed of ⅝-in. end plates and ⅜-in. side and web plates and with angle-iron joints. The angle-flanges are 4 by 4 by 1½ in. thick. The stacker-ladder is 132 ft. long between centres of drums and the conveyor-belt is 32 in. wide, ⅝-in. thick and 174 ft. long.

The ten dredges working in this district are digging the actual bottom and are co-operating with the Debris Commission in building their dams. By referring to the acompanying sketch map (Fig. 100) an idea of this work may be obtained.

The Yuba bottom is shown by the enclosing heavy lines, which represent approximately either the old levees or natural high banks, and in winter or flood-time the area enclosed is almost completely covered with water for its width of nearly two miles. In summer, the flow is represented by several streams that, constantly changing their course, leave bars and quicksands where, within a few hours, the current was racing. The main portion of the river flows at present as shown in the sketch, but it is to be diverted through the cut behind Daguerre Point recently completed by steam-shovel. The dam, which is thoroughly described in the MINING AND SCIENTIFIC PRESS of September 2, 1905, is situated about three miles

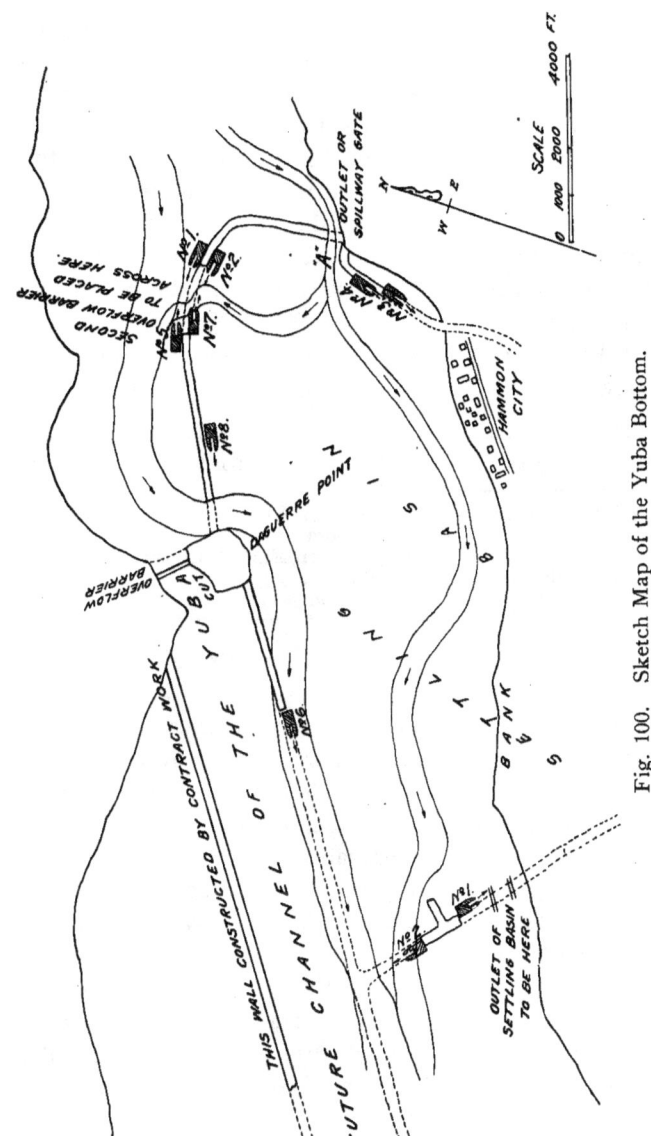

Fig. 100. Sketch Map of the Yuba Bottom.

above the present dredging ground. The training walls or dams, shown by double continuous and dotted lines, are being built by the dredges and consist of the stack piles. The cut behind Daguerre Point and the north training wall have been already built, as far as shown, by the Government; and the dredges of both companies are working in the directions shown by the arrows. At A a spillway or inlet is to be made; this can be raised or lowered, so that the settling basin in the low ground behind Hammon City may gradually be filled. At B is the outlet to this settling basin, which receives the overflow east of the east training wall and will have a capacity of 5,000,000 cubic yards.

At Folsom there are three companies operating seven boats, as follows:

Name of Company	No. of Boats	Builders.	Capacity of buckets in cu. ft.
Folsom Development Co.	2	Western Eng. & Construction Co.	5.8
" " "	2	Folsom M. Co.	8.9 and 13
" " "	1	Western Eng. & Construction Co.	8.5
Ashburton Mining Co....	1	Bucyrus	9
Eldorado Dredging Co. .	1	Risdon	6

There are four other old boats in the district, one of which might be remodeled and used.

The dredging ground along the American river is composed of successive benches and the bars formed by the old courses of the river. The sketch (Fig. 101½) is taken from 'Report of the California Bureau of Mines' and shows how the course of the river has changed. There are probably some 6000 acres in all, which may be dredged, and though some of this is of low value per yard this portion is generally soft and easy to dig. The costs in this district are low.

No. 4 boat, belonging to the Folsom Development Co., is the largest dredge in the world and the following outline description will give an idea of its size and equipment. The design of the dredge is a combination of Bucyrus and Folsom Development Co.'s shops, and was erected by the latter. The hull is 102 ft. long and 58 ft. wide. There are 59 buckets, each of 13 cu. ft. capacity and the digging-ladder is 66 ft. 11 in. long. Both the upper and lower tumblers are hexagonal. The dimensions of the shaking screens are: Upper screen, 10 ft. 6 in. by 11 ft.

Fig. 101. The Treacherous Yuba Bottom. Here it is Two Miles Wide.

11 in. Lower screen, 14 ft. by 12 ft. 8 in., and together they contain a total superficial area of 300 sq. ft. The Holmes system of gold-saving tables is used. The stacker ladder is 85 ft. 6 in. long and the conveyor-belt is 44 in. wide and sets of four idlers are used.

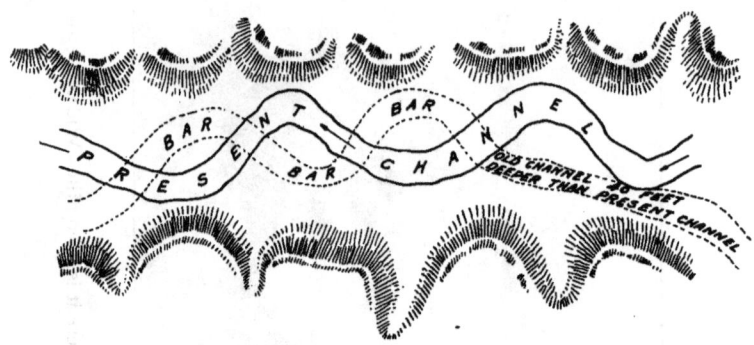

MIDDLE FORK OF AMERICAN RIVER.

Fig. 101½.

The motors are as follows:

Name of motor.	H. P.	Description.		
Ladder hoist and main drive	150	Westinghouse	variable	speed.
Main pump	100	"	constant	"
Auxiliary pump	15	"	"	"
Screen drive	30	"	"	"
Stacker	30	"	"	"
Headline and sidelines	30	"	variable	"

The boat is worked entirely on head line.

IX. APPENDIX.

Gold Dredging.

The Editor:

Sir—I have had occasion recently to inquire into various aspects of gold dredging. I find that whereas the purely technical side of this industry, such as dredge designing and operating, has been written about a good deal, the statistical and financial aspects have hardly been dealt with at all.

Let me enumerate some questions to which as yet I have not found satisfactory answers:

1. How many payable, or probably payable, dredging areas are known in the world, and what is, approximately, their gross extent? Which are the likeliest countries to prospect?

2. How much gold is produced yearly by dredges, and how much of the production is profit?

3. How many dredges are now working at a profit?

4. How many dredges that were once worked are not now working, and what were the reasons for their stoppage? How much money do these represent?

5. What are the precise costs in different localities? How are costs affected by
   a. Type of dredge?
   b. Nature of deposit?
   c. Nature of power?
   d. Climate?

6. What is the best system of prospecting dredging ground? How should such system be adapted to different kinds of ground? Is drilling as effective as sinking shafts or paddocks? What is an average recovery of the gold contents as shown in prospecting? Why are favorable results in prospecting often falsified in practice?

7. Assuming prospected or otherwise proved ground to represent 'ore blocked out', which is the correct basis of valuation for such ground?

If gold dredging is to be regarded as a scientific industry, it seems to me that satisfactory answers to such questions as these should be procurable. A mass of facts must be available by now. Can you not, through your many sources of information, undertake the collection of these?

The cardinal fact which would seem to emerge from my inquiries is that the known areas of profitable dredging ground are not numerous, and are limited to a few countries. When recently in San Francisco I had this statement indorsed by the man who, I consider, knows more about gold dredging than anyone else. He has for years past sent experienced prospectors over the United States, South America, the East, and elsewhere, but, I understand, outside of his Californian properties has found nothing definitely good. Of course, one must not jump to the conclusion that no more good areas will be found. South America, Alaska, Siberia, India, and the Eastern islands still offer an immense field for research; and, although ideal dredging conditions cannot exist in these countries, there is no reason why they should not contain many payable areas. In the meantime, it may be taken for granted that the definitely good areas are limited. Ideal dredging conditions exist, I believe, only in certain parts of California and New Zealand, while Victoria ranks next, but a long way behind. At present the two first-named countries are probably producing 75 per cent of the gold won by this method. Dredging in the United States is, I understand, almost confined to California. I make the statement, subject to correction, that there are not more than four dredges working elsewhere at a profit, and that outside California conditions for working are usually unsatisfactory. Several dredges are said to be working at a profit in Montana, but I have found no one who can give me exact statements. In California, the chief centre of dredging is Oroville, where the conditions are quite ideal. A recent visit to this field impressed me with the idea that the Oroville gravels—proved as they are—form the safest gold mining venture in the world. At Folsom there are also good dredging areas, and Bear River should soon become profitable. But the centre of interest in gold dredging will soon shift to the Yuba river. Here a large area of gravel 90 ft. deep has been found to give good values, and several dredges already at work have confirmed these results. One company, alone, will shortly be working 15 dredges, of a new and very powerful type. The strength and efficiency of these is such that they will handle 90 ft. of gravel as easily as the other types can handle 30 feet.

In Canada, dredging on the Fraser and Saskatchewan rivers has been, somewhat unaccountably, a failure; I should like to know the reasons for this. I should also like to have precise figures about

APPENDIX. 185

the dredges working at Atlin, British Columbia. It is stated, vaguely, that they are working profitably, but, knowing Atlin as I do, I am skeptical. This same frame of mind clutches me when estimating the chances of dredging in all the northern regions—Klondike, Alaska, Siberia. The climatic factor in these countries seems to me almost to exclude the hope of profitable results. The short season, the perennially frozen condition of much of the ground, the heavy cost of labor, and the great distance from foundries and sources of supply, are surely drawbacks of almost vital importance. I am told that dredges are now working profitably both in Klondike and Alaska, while the Russian Minister of Mines assured me that a number of dredges were getting payable results in Siberia. But I am still skeptical. I have never seen figures or results from any of these places. I shall want to see the balance sheets for, say, two consecutive years' work by some of these dredges, before I can believe that in these countries the climatic factor is not overpowering.

South America, with its immense river systems, offers a great field for possible dredging areas. As yet, results there are of doubtful value. Colombia has been a good deal prospected, with, I believe, not satisfactory results. In Dutch and French Guiana some dredges are working, but I do not know with what success. I lately read a report on ground in Peru, stated to go 75c. per yd. The writer was a man of presumable experience; had he not been, I should have looked on such a statement as a fairy tale. This ground lies on the eastern side of the Andes, and a dredge would have to be transported in sections. Moreover, as for a long distance there is a precipitous descent, and that by bridle-track only, the sections must be no heavier than a couple of mules could carry between them. It seems to me that a dredge thus designed, especially if working in heavy ground, would surely be a failure. I should like information on this point—not only as to whether it can be sectionalized down to, say, eight cut pieces—but as to whether it can be expected to do satisfactory work on that basis. Buenos Ayres has recently been the scene of a dredging boom—on paper. The locality of the area thus boomed was the river San Juan de Oro, in Bolivia. The first dredge started there worked for six weeks without getting as much as one ounce of gold. It then transpired that the men in charge—picked experts of reputation from New Zealand—had been badly 'salted'. I am afraid there is not much chance for success in this locality.

One of the first dredges worked in South America was located in Matto Grosso, the interior province of Brazil. This dredge has demonstrated the existence of gold there; and the reports of prospectors, if they can be relied upon, point to a large area of possibly payable ground. It is notable that the Matto Grosso gravels carry also diamonds, which are caught on the dredge in considerable numbers. Brains are now at work devising a supplementary plant, which will recover the diamonds after the gravel has been dredged for its gold. Elsewhere in Brazil there is a good deal of prospecting being done, but in a manner I would describe as superficial. I saw lately a dredge at work in the river near the old mining town of St. John Del Rey. This belonged to a New Zealand company, and, I believe, was paying expenses. In Tierra del Fuego there are large areas of gravel carrying gold. Those on the Argentine side were extensively prospected by two friends of mine last season. The conditions were pronounced as favorable, but there was too little gold present. On the Chilean side there are some undoubtedly rich areas; but here the ground is rougher, and would require a specially adapted type of dredge. These areas may, I think, be reckoned as possibilities.

New Zealand, Victoria, and New South Wales, alone among Australasian States, seem to contain payable gold-dredging areas—though in Tasmania dredging for tin is now being done. Even about dredging results in New Zealand—the home of the industry—there are surprisingly few facts available. I believe that one-third or even one-half of the dredges started in New Zealand have worked at a loss; this would seem to indicate that the methods of prospecting there are casual, and that there are no systematized results to guide would-be valuers of ground. I am told that all possible dredging areas in New Zealand have long ago been secured, also that at the present rate of exhaustion most of the best ground will be worked out well inside 12 or 15 years.

After numerous failures, especially adapted dredges, or, in cases, hydraulic pumps, are now securing fair aggregate results in Victoria and New South Wales. Most of the areas worked, however, are of small extent, and profits are on quite a small scale. As elsewhere, it seems likely that the areas in Australia where dredging conditions are favorable will prove to be limited.

In Rhodesia, or rather over the Rhodesian border, in Mozambique, there was an abortive effort at dredging, the conditions proving unsuitable. I believe this is the only dredge erected in South Africa. In West Africa there is certainly gold in the rivers, but I am skeptical as to general conditions. There are some half dozen dredges now at work there, but their results are indifferent; moreover, I imagine that the richest patches are being worked first.

As regards the East, there is not much yet known of its dredging areas. In Burma, on the Irrawaddy, a prospecting dredge has been at work for some time. I understand that it has shown profitable figures, but I have no facts about this. However, I look on Burma as among the possibilities. It seems to me, also, that India is a country to be prospected, but I know of nothing as having yet been done there. Two dredges are at work in the Malay State of Kelantau; from the published returns I judge that they are not paying. In Korea, the Oriental Company is to try an experiment in dredging; while it is stated that in Borneo and the Celebes there are hopeful prospects.

In Europe, there are several dredges at work in Servia. I believe these have earned a commercial profit, for others are being built; but, as usual, there are no precise figures available. In Hungary, there might be dredging areas, but I suppose there is little to be expected of the rest of Europe.

As I started out to ask for information, not to give it, I think I have said all that is compatible with my premises, and I shall not venture into the details of a discussion on which so many others must be better informed. I can, however, say first this, that I believe the Oroville 'school' of dredging, both in methods of prospecting and in designing, is now quite ahead of New Zealand, and that the work now being done at Oroville and Yuba river should be studied by all who wish to master the details of this subject. The exact results from the Oroville field to date, together with the description of the prospecting methods employed there, and an account of the evolution in the type of dredge now in use there, written by those who know, would form a valuable nucleus of facts on which to base a general inquiry into the whole subject.

J. H. CURLE.

St. Moritz, Switzerland, December 24, 1905.

Gold Dredging.

The comprehensive glance at the present state of the dredging industry, which we published in our last issue, will have interested many of our readers. Few men have such acquaintance with gold mining regions as the author of 'The Gold Mines of the World', and therefore Mr. Curle's letter commands respect. It was a useful communication, for not only did the writer of it give information, but he fulfilled the equally useful function of inciting donations of knowledge from others. Certainly the knowledge available on gold dredging is amorphous and incomplete; the only published book upon it is rubbish, because it was compiled by a man of no special experience. The best publication on the subject is the bulletin issued last year by our State Mining Bureau. Most of the articles heretofore written have been too vague, too little anchored to facts and figures; a few practical contributions emerge out of this mass of hazy literature and these have been prepared, almost without exception, by Western engineers. But there is much to be done before the gaps in our information on this important branch of mining are filled. Mr. Curle specifies the several points concerning which we need facts and figures, and we are confident that among our readers there are men able and willing to contribute what is wanted. Already we are enabled to publish a special article on the testing of dredging ground, with a workmanlike description of the methods to be used in checking such results.

As to prospecting ground with drills, it is said that the Oroville experience has shown a yield by dredging from 70 to 85 per cent of that given by drilling tests. It is just such statements as this that are useless, although apparently businesslike. The value of a series of drill-holes as indicating the richness of a tract of gravel cannot be gauged by any empirical formula. It is as absurd as the practice of timid engineers who cut their estimates of ore in two or deduct a certain fixed percentage from their calculations, so as to be safe. In the case of drilling before dredging, the result is reliable according to the number of holes, the distribution of them, the care taken in the work and, above everything, the personal factor. Such work carefully done and checked at each stage by a man of experience and integrity needs no big

APPENDIX.

discounting, while that accomplished by an unreliable driller or a careless novice is worse than worthless. No fixed percentage covers the varying conditions surrounding an engineering operation.

In the early days of dredging it was the custom to test 200 acres with 10 drill-holes; nowadays one hole to two acres is not uncommon. When results indicate that the gravel is of varying depth, that there are irregular channels or that the ground is spotty, it is not unusual to put down as many as one to three holes per acre, especially where no adjoining workings exist, such as throw light on any anomalous results from the drilling. The cost is a limiting factor, for each hole costs $60 to $100 in ground of any considerable depth.

It is generally assumed that the results given by shaft-sinking are more reliable than those from drill-holes; as a rule this is true, but the comparison too often smacks of the old idea that a mill-run is more trustworthy than the sampling of ore in a mine. It depends upon how it is done, with this proviso, that the more men needed to carry out an operation of sampling, the greater the opportunity for error, intentional or unintentional. A shaft gives better opportunity for examining the nature of the successive layers of gravel and other conditions that bear upon the subsequent working of the ground. Careful measurements are imperative; sometimes it is not practicable to hold the same diameter of shaft all the way down, and in running ground one has to resort to timbering; these factors affect the cross-sectional area and must be carefully noted in any calculations. Moreover, instances are known where shaft results have been seriously vitiated by the fact that the particles of gold have fallen to the bottom, with the water, so that the upper layers of gravel appeared worthless, while that immediately above bedrock was excessively enriched. Great caution, bred of experience, is required to make an accurate test.

Editorial MINING AND SCIENTIFIC PRESS, February 3, 1906.

Sectional Dredging Machinery.

In our last issue, we endeavored to stimulate a discussion on gold dredging by referring to the methods employed in testing ground by drilling. This week we publish the continuation of an article on this subject. Another important point in connection with dredging is the degree to which it is safe to sec-

Fig. 102. Wreck of the Colorado-Pacific, one of Earliest Boats Built at Folsom.

tionalize the heavy parts of a dredge. It is obvious that such subdivision must be a last resort. To sectionalize dredging machinery, the chief essential of which is strength, is like dividing a stamp-mill mortar where solidity is the first requisite. The ordinary dredge equipped with 5-ft. buckets employs a shaft and tumbler weighing together about six tons; the shaft, made of nickel steel, is 14 inches in diameter and is 8½ feet long, so that it weighs over a ton—usually about 2500 pounds. Each bucket-base weighs 1100 pounds, and even the pins are over a hundred pounds apiece. Obviously, to cut down the parts of such machinery to the 250 or 350 pounds that a mule can carry, is a difficult matter. The tendency in California is to attain efficiency and economy by building the dredges enormously strong and big. There is a dredge at Folsom—treating 150,000 cubic yards per month—that has close-connected buckets of 13-ft. capacity, and another 13-ft. dredge, but of the intermediate or linked bucket type, designed by Julius Baier, has been erected in Montana. Dredges such as these are beyond the transporting capacity of all the mules in creation. The ordinary and, as regarded nowadays, small machine with 3-ft. close-connected buckets is provided with an 8-inch shaft weighing 1155 pounds and uses a bucket-base of 475 pounds. Even such a dredge could not be transported over mountainous country. To decrease the burden, the shaft could be worked into the tumbler casting, so that the weight would be distributed into thirds. The bucket-base is difficult to reduce, save by cutting in half, at big expense. This applies to the close-connected type; for a bucket and link dredge of 3-ft. capacity, a six-inch shaft would suffice and the weight would be one-half that given above. By using a built-up bucket-base instead of a solid steel casting each piece can be reduced to 150 pounds; with sacrifice of strength, of course. Such construction will require .a pneumatic riveting machine for assembling the parts. However, these are details beyond our province. We desire merely to stimulate intelligent discussion on the subject and we trust our professional friends will avail themselves of the opportunity.

Editorial MINING AND SCIENTIFIC PRESS, February 10, 1906.

Sectional Dredging Machinery.

The Editor:

Sir—As regards sectionalizing placer dredging machinery for use in remote regions; it is obvious that the boilers, engines, and such parts as have been sectionalized for other purposes may be readily obtained in shape for mule-back transportation. There are, however, various portions of the machine which it seems at present could not be sectionalized satisfactorily to the required weight or size of piece which may be transported in this manner.

The standard type of dredge used in California uses a bucket having a capacity of 5 cu. ft.; it has a cast-steel back weighing from 850 to 1000 lb. Bucket-backs of lighter weight have been discarded in favor of the heavier, on account of the increased wear obtainable. It is true that a built-up bucket either of the continuous or open-link type can be built of small pieces riveted together, but actual practice demonstrates that the solid cast-steel back is preferable in every respect. It is true we are using in California some dredges with 3-cu. ft. buckets, those of the open-link type with built-up buckets having an actual capacity of from 600 to 750 cu. yd. per day, while the continuous-bucket type with cast-steel bucket-back will average from 1000 to 1500 cu. yd. per day, in the average digging, without trouble. The upper tumbler for driving the bucket-chain it would be practically impossible to sectionalize; in fact, the latest practice in this respect calls for a six-sided tumbler, in place of the old five-sided, not because six sides are better for driving or for the dumping of the bucket, the fact being to the contrary, but because practice has established a size of shaft which it is practically impossible to pass through the old type of five-sided tumbler. The shafts are getting heavier annually. From my personal experience, dating back to 1893, I can say, that at that time we considered a $6\frac{1}{2}$ to 7-in. shaft as ample for driving a 5-cu. ft. bucket-chain into the tightest of soil. Recent construction calls for hollow-forged and oil-tempered nickel-steel shafts, the very finest that can be produced, of a diameter ranging from $11\frac{1}{2}$ to 14 in.; and as these shafts are from 8 to 10 ft. long, according to the type of drive used, the piece is necessarily very heavy, and there seems to be no possible way to reduce its weight for packing over a mountain trail. In fact, one five-foot dredge, to my knowledge, was fitted with a hollow-steel shaft 18½ in. diam., and approximately 11 ft. long, which it would have been impossible to have made in more than one piece. Other various portions of the dredge could probably be sectionalized to quite an extent, and in the case of remote operations a dredge could undoubtedly be built which would handle a reasonable quantity of gravel and not have any parts too heavy for mule-back transportation. A dredge which closely approximates the necessary conditions for this purpose was built by Mr. Ogilvie, of Dawson, from designs of A. Wells Robinson, of Montreal, and was put in operation on the Stewart river. Buckets were of 2¼ cu. ft. on alternate links of the chain, and the dredge was intended for excavation not to exceed 15 ft. below water-line. The dredge has been successfully used in prospecting various portions of the Stewart river, and now lies on the Yukon a short distance below Dawson. It will probably be used next season for prospecting other parts. The fact still remains, however, that if one has sufficient acreage, of placer ground which is suitable for dredging purposes, and which is remote from the common carriers, if the ground is sufficiently rich to warrant his dredging it for a profit, the cost of constructing a road which would permit his machinery to be hauled to the ground in the usual form would be so small in proportion to the profits on the operation that distance from transportation need not deter anyone from entering into the dredge business. I have certain knowledge of machines of the usual type which have been hauled over mountain trails 70 to 150 miles from a railroad station, and in cases where the owner was obliged to make the road or trail over which the machinery was hauled, and in one case even to send to the Eastern States for wagons, harness, etc., and to Kansas for mules, in order to get his machinery on the ground, and the dredge has repaid its entire cost of installation, and all expenses connected with purchasing the ground, prospecting it, and maintaining it, and a profit besides of over 150% on the total capitalization of the venture, although it has only been in operation three years. The prospecting shows an average value of 16c. per cubic yard. This dredge is driven by steam and burns the poorest quality of spruce wood for fuel at a cost of nearly $4 per cord at the dredge, and the dredge consumes 15 cords per day.

G. L. HOLMES.

San Francisco, February 8, 1906.

From MINING AND SCIENTIFIC PRESS, February 17, 1906.

Gold Dredging.

The Editor:

Sir—Mr. Curle's criticism, in the issue of January 27, 1906, of the dredging industry is timely, and worthy of the attention which he evidently wishes it to receive. I will give a little consideration to some of his points, with reference to the Alaska field, and other regions.

In order to consider this question: Divide, mentally, Alaska into three provinces, (a) South Coast province, (b) Interior province, (c) Seward Peninsula province. Now eliminate all permanently frozen ground. This consideration strikes out at once almost all of the Seward Peninsula, including the region about Nome, and all north of Nome to the west of the treeless area. It also bars the whole of the Interior province, with the possible exception of extraordinarily rich creek-beds, like Bonanza creek on the Klondike, where the material has been already loosened and disturbed by previous work, and portions of the Klondike river-bed. The South Coast province only remains. I would, therefore, eliminate from consideration as dredging ground at least four-fifths of the Territory of Alaska.

By the South Coast province I mean the strip extending from the southern boundary, along the coast, north and west; this includes the high frontal ranges of mountains of the Alexander archipelago, the St. Elias and Yakutat districts, Kodiak island, the Kenai and Alaska peninsulas, the Aleutian islands, and the area lying between the north shore of Bristol bay and the Kuskokwim river. In the areas mentioned permanently frozen gravel does not occur.

All the territory lying east of the 149th meridian may, I think, be eliminated. Flood plains reduced nearly to base level, in which streams meander, are in my opinion the first requisite for auriferous gravels fit for dredging. Unless this prime geological feature is present, it is not likely that suitable ground exists, and the engineer wastes his time and that of his client in examining the area. As I do not own dredges, and further as I do not have cognizance of any balance sheets except those of the California companies, it is impossible for me to answer some of Mr. Curle's inquiries. I believe, however, that in Alaska there has been no successful operation up to the present time in the South Coast province. It is reported that

APPENDIX. 195

a dredge built on the Solomon river, Seward Peninsula, which has operated during the season of 1905-6, has been successful. We have no published statement about it up to the present, however. Granting the success of this dredge, it is due to exceptional, very exceptional, conditions in that district and latitude, a repetition of which one may not expect to encounter. I therefore unhesitatingly condemn the Seward Peninsula as a dredging province. By this I do not wish to say that rich placers do not occur there. The gold is there, but among the methods of working, dredging is not to be considered.

The portion of the South Coast province which appears most promising for dredging lies to the west of Cook inlet, the Sushitna river and to the south of Bristol bay. The tributaries of the Yentna river, which is itself a tributary of the Sushitna, should be investigated. Outside of the area above named, I know of no Alaska area which is worth the dredge-investor's prospective consideration. Mr. Curle asks: How many payable dredging areas are known in the world? I should say portions of California, including the Feather and Yuba river areas, and parts of New Zealand. As regards probable areas, I may be permitted to suggest eastern Russia. I witnessed dredge construction there in 1902. I have heard that several dredges have since been constructed. The factor of climate, to which Mr. Curle refers, needs some definition. If he refers to permanent frost in the gravel, I am in entire agreement with him, and would offer the same objections to the Olekma, Vitim, and Amgoon districts as to the Nome district. If, on the other hand, he refers to the disadvantage of short working season to which northern areas are subject, I do not think his objections altogether valid, as I believe the Ural offers an extensive and promising field for exploration.

The Atlin district is, I will admit, an extremely hazardous locality for dredging. The gravel is not particularly rich. By this I mean that the ancient benches, if worked as a whole (and there is no other auriferous alluvium to be considered there), will not run over 20c. per cu. yd. The wash is extremely heavy, and the thickness is above 90 ft. except in very limited areas. The season does not exceed 150 days, and even this is long for that region. For the matter of gold and its profitable exploitation by some form, Atlin is not without possibilities, but the method will have to be some other than dredging.

Dredging in South America is a fact, but the conditions of access to most of these regions have hitherto been prohibitive, and it may be said that dredging capital will turn its attention to many other parts of the world before hazarding investment in South America. Recently we have heard of a gold and diamond dredging field in the vicinity of Diamantina, Brazil. The ground is now being explored and I understand the gravel is being drilled. It is possible there is a profitable field for dredging investment in this isolated region, but it will be some time before results can be given to the mining public.

Mr. Curle says that probably one-third, or even one-half, the dredges started in New Zealand have worked at a loss. This fact is not a condemnation of a dredging province. In California, outside the comparatively small area about Oroville, Marysville, and Folsom, there are many so-called dredging districts, but the investigator finds that the word dredge is much more common than the word dividend.

In the United States proper, with the exception of the operations already noted in California, the public knows of no profitable gold dredging, from authenticated published statements, although in Montana, Idaho, Arizona, and Colorado dredging operations have been attempted, and there are rumors that some dredges have been made to pay. Georgia has furnished one small patch of dredging ground which I believe was payable, namely, that of the Chestatee river, below Dahlonega. I do not think it impossible that limited portions of river beds in auriferous areas of North and South Carolina and Georgia may yet be profitably dredged. It is highly probable that in Burma and British India dredging fields will yet be found. The peculiar conditions of weathering which exist there in the auriferous belts should afford the proper conditions for soft bedrock, and the portions of the river flood-plains, lying midway between the sources and the sea, should be prospected.

In regard to the amount of gold which is produced yearly by dredges, Mr. Aubury has told us from reliable data how much is produced in California, and I believe this represents nearly the whole product of the United States. The amount produced in New Zealand I have no means at present of knowing; the amount produced in other parts of the world is at present insignificant. The estimate of future production in California is conservatively $200,000,000, as may be deduced from Mr. Aubury's figures.

I should say that the number of dredges in all parts of the world which are working at a profit does not exceed one hundred. Of the number of dredges which were once worked and are now not working, one can form only the roughest estimate. If one hundred are working at a profit, it is safe to say there are five times this number idle or abandoned. Although this seems a large penalty to pay for experimental work, if the expenditure eventually results in the establishment of a sound producing business in various parts of the world, it will not be considered excessive.

The precise costs of dredging in various localities are probably known to Mr. Curle as well as to anyone. Eight cents at Oroville, 80c. in the Klondike, and 50c. per cu. yd. in the Seward Peninsula of Alaska, are the figures of which I feel fairly sure. The high cost in the Klondike is where gravel is thawed by steam-points in front of the dredge as it progresses. The Alaska figure assumes a season of 120 days, and delays are occasioned by annual, but not permanent, frost. New Zealand costs are said to be very low, but I have no figures at hand.

Regarding costs in the Ural region, which I have indicated as the best field in the Russian empire, all conditions considered, I would allow an average season of 180 days. The cost on this basis, assuming a ten-year life, not to exceed $500 per acre for the ground as purchased, $75,000 each for the dredge installations with a capacity of 1500 cu. yd. per day, in 30-ft. average ground, is 14c. per cu. yd. This is taking the highest figure for operating at Oroville and adding to it depreciation, amortization, and interest on total invested capital, making an extra charge of six cents per cubic yard.

In general, the cost of dredging is not seriously affected by nature of deposit for the reason that dredging in deep ground, for example 60 ft., merely increases the first cost of installation and does not seriously affect the operating cost. Ground containing numerous and large boulders cannot be dredged at all, so this kind of deposit need not be considered. Likewise ground with too great an overburden of turf or muck cannot be profitably dredged. Likewise ground which is too shallow cannot be dredged, but should be worked by steam-scrapers. By elimination of the objectionable types of ground, it will be found that the cost of working real dredging ground will not greatly vary in California, Burma, or European Russia. If anything, the operating cost will be somewhat greater in the two latter countries. As regards power, where dredges are operated, as at Oroville, by a general power-system, the cost will be less than for self-contained dredges operated by steam, unless fuel is remarkably cheap.

Prospecting of dredging ground should be done by pit-sinking if possible; as this method is rarely feasible, drilling with churn-drills is generally requisite. This subject has been so thoroughly discussed recently that I shall leave it to those best qualified to answer. For the determination of the proper factor to apply in each case of an examination, I unhesitatingly recommend the sinking of a pit around the drill-hole if possible, in at least ten per cent of the number of holes drilled, to check the drilling results.

The proper valuation of dredging ground should be made on a system which eliminates all doubtful territory from consideration. For example, in a base-level valley suitable for dredging there may be an average width of 1500 ft. of gravel-flat covered by marsh, forest, underbrush; in any of which an open-cut to accommodate the dredge-hull may be dug. There may be a meandering stream four or five feet deep, which is subject to spring freshets occupying 500 ft. of a given cross-section of the valley. The danger of working in this stream should be taken into consideration, as the dredge may be wrecked by a flood, if operations are attempted in the present water-course. Let us say that all the area outside the limits of the river is workable. But at best, only a small area near either bank in the river itself is dredgeable, and it is a risk to allow for the gold which may underlie the main body of the present stream. To this it may be objected that in certain dredging areas the auriferous channel is coincident with the course of the present river. I reply that an area presenting such geological conditions is not in a bona fide dredging province, as it does not represent dredging ground founded on the Oroville province as a type; namely, where the gold is fine and distributed with a fair amount of uniformity over a wide comparatively shallow area.

In further considering the valuation of dredging ground, I would urge the suggestion that while operating expenses have been published in considerable detail, the valuer of dredging ground has few data concerning the cost of ground and equipment. If operating expenses at Oroville, for example, are being reduced to five cents per cubic yard and under, what additional cost per cubic yard must be figured for the annual consumption of the ground, and to allow for the amortization of the machinery? Scrap iron is no great asset,

APPENDIX. 199

and that is all a dredge amounts to after the gold which it is winning for its owner is removed.

I think it must be confessed to inquirers like Mr. Curle, that the proper system of valuing ground on such a basis as to show the probable net proceeds of the operation, has not been discovered. At least I have never seen published a balance sheet of a gold-dredge which shows how much it earns over all expenses. Oroville has given us the data for estimating actual as against theoretical work in cubic yards handled. The present excessive cost of, and time consumed in, repairs will probably be reduced, but Oroville experience gives us the best basis for estimating, as yet. The life of the property can be determined by prospecting, and the yardage and cost of each dredge being a fairly well-determined factor, operating cost being estimated from local conditions, the eventual net proceeds ought to be foretold with a fair degree of accuracy. No valuation of dredging ground which does not thus allow for amortization of the total preliminary expense of acquiring, prospecting, and equipping, in addition to fixed charges for operating for a given length of time, appears to me worthy of consideration.

C. W. PURINGTON.

Denver, July 15, 1906.

From MINING AND SCIENTIFIC PRESS, July 28, 1906.

Gold Dredging.

The Editor:

Sir—In the issue of your paper of January 27, 1906, under the heading of 'Gold Dredging,' J. H. Curle asks seven specific questions, and the article goes on to give a great deal of valuable information which, on two points, I venture to think, is a little misleading. In referring to the Yuba district, Mr. Curle states that a large area of gravel 90 ft. deep gives good returns, and that the dredges there are to work to that depth. Though the drilling in the Yuba field discloses the fact that gold values are contained at least to a depth of 115 ft., it has only been considered advisable to build boats fitted to dig at their greatest advantage to depths not over 70 feet.

This article was published in January, 1906, and at the present date there are ten dredges at work. All of these are equipped to dig to 64 ft. below water line, and the deepest work that is, or has been, done there is at a depth of 72 ft. below water line. Under the conditions prevailing on the Yuba the water line is, as a rule, only a few feet below the actual surface—in some cases, indeed, the dredging ponds have to be kept pumped out so that the surface of the dredging ground may be above water and in order that the digging ladder may reach the depth mentioned. This, so far as I am aware, is the greatest depth to which dredging has been carried in any part of the world, but as the best values in dredging lands in the majority of cases lie close to the bottom of the digging it is not hard to see that an overburden, if commercially unproductive, simply tends to reduce the profit—or, in other words, to lower the average grade per yard of the material which it is necessary to handle.

The other point to which exception may be taken is contained in the concluding sentences where it is inferred that Oroville and Yuba districts should be studied as the most modern examples of up-to-date prospecting, designing, and dredging. The name of Folsom is only conspicuous by its absence. In my humble opinion this district should rank at least with the others for all and more of the qualifications mentioned. It is not only the most important as having the largest available area of dredg-

APPENDIX.

ing ground; but (in some cases under unique conditions) illustrates the handling of the largest yardage per boat of any locality in the world, and—as far as can be learned—at as low an average cost per yard.

I have recently completed an extensive examination of the dredging fields of Oroville, Yuba, and Folsom in the Sacramento valley, and have investigated fully most of the points embodied in Mr. Curle's seven specific questions. Those questions were:

1. How many payable or probably payable dredging areas are known in the world, and what is approximately their gross extent? Which are the likeliest countries to prospect?

2. How much gold is produced yearly by dredges and how much of the production is profit?

3. How many dredges are now working at a profit?

4. How many dredges that were once worked are not now worked, and what were the reasons for their stoppage? How much money do these represent?

5. What are the precise costs in different localities? How are costs affected by: *a.* Type of dredge; *b.* nature of deposit; *c.* nature of power, and *d.* climate?

6. What is the best system of prospecting dredging ground? How should such system be adapted to different kinds of ground? Is drilling as effective as sinking shafts or paddocks? What is the average recovery of the gold contents as shown in prospecting? Why are favorable results in prospecting often falsified in practice?

7. Assuming prospected or otherwise proved ground to represent 'ore blocked out,' which is the correct basis of valuation for such ground?

As the results of my investigations are shortly to appear in print, I shall simply refer to Mr. Curle's queries in a brief manner.

No. 1 does not apply to this locality.

No. 2. This is a difficult question to answer accurately in detail, because the information is not fully obtainable. However, assuming that at Oroville, the average total working cost is 7c. per yd., the average recoverable value of the gravel 14c., and the annual yardage produced about 19,200,000; then there is a total annual production of $2,688,000, of which $1,344,000 is net profit. On the Yuba the annual yardage is from 9,600,000 to 10,200,000. The recoverable value of the gravel has been estimated at 20c. per yd., then assuming that the working cost is 5.5c per yd., the total annual production is $2,000,000, and the profit $1,450,000.

At Folsom figures of this sort are difficult to obtain, but the average value over the whole tract is probably not over 11c., though in places it will run much higher than this figure. Taking the average working cost as 5c. per cu. yd., and the total annual product as 8,500,000 cu. yd., the profit is estimated at $510,000.

Therefore in the three districts an approximate estimate of the total annual production, the average recovery per yard, the average cost of working per yard, and the total annual net profit may be summarily expressed in tabular form as follows:

District	Annual operations in cu. yd.	Production	Recovery per cu. yd. in cents	Cost per cu. yd. in cents	Annual profit
Oroville	19,200,000	$2,688,000	14	7	$1,344,000
Yuba	9,600,000	2,000,000	20	5.5	1,450,000
Folsom	8,500,000	935,000	11	5	510,000
Total	37,300,000	$5,623,000			$3,304,000

No. 3. In the three districts there are forty-nine dredges in operation, of which, say forty-one have been working over nine months. Of the latter number probably over 90% are working at a profit, while undoubtedly all of the remaining eight will pay dividends.

No. 4. Only two dredges that are not now working are still fit for operation, all the others having, through service, become mechanically unfit for use. Of these two, one is of too small capacity and is practically worn out, while the other is also of ancient design, but could be remodeled and used. After litigation it was sold at sheriff's sale. The property on which it operated is still commercially valuable from a dredging point of view. The two dredges probably originally represented about $95,000.

No. 5. The costs have been given in the answer to question No. 2.

a. The type of dredge affects the cost of production chiefly as regards its digging capacity, this being favorable in proportion to the increase in size of buckets up to at least 7½ cu. ft., and probably up to 13 cu. ft., where proper repair facilities are avail-

APPENDIX. 203

able and several machines are operated under one management. Fixed costs are reduced in the larger machines and output is increased.

b. The nature of the deposits affects the costs as follows: Hard ground and large boulders require more power and there is greater loss of time, one of the main factors in increasing cost. Moreover, the cost in labor and repair of parts is increased.

Clayey ground requires more disintegration and causes loss of gold over the stacker.

c. Electric power is the only form used in the districts of Oroville, Yuba, and Folsom, and is undoubtedly superior to steam, not only as affects reduction of cost, but on account of its advantageous application and control.

d. Climate practically does not enter into consideration in this region except as regards high or low water in the flood season. Even then it does not affect the cost to any degree.

No. 6. Undoubtedly the most effective practical system of prospecting dredging ground is by shaft. Drilling costs on the average $2.50 per foot, and is often very unsatisfactory in results, both as regards information on values and nature of ground. In comparison with the drill as a method of sampling, a shaft is more thorough and costs less per foot above water-level; even below water-level, where pumping is necessary, it is to be recommended, as it can be referred to afterward in ascertaining the exact nature of the ground. The cost of shaft sinking by the 'China' method varies between 90c. and $1.50 to water-level, and the increase below this point varies with the amount of water. These figures include cost of panning.

The ratio between estimated value of ground by drilling, and actual recovery by the dredge, is not a fixed proportion, but may generally be said to increase in constancy over a large area roughly in proportion to the number of holes per acre. Single holes are rarely indicative of the value of adjacent ground, though a somewhat arbitrary rule that seems to apply in practice, is: Where a hole gives excessively high results the recovery by dredge at that spot is usually lower, and vice versa.

The care with which the computation of the value of a dredging property from bore-holes is made should be in exact proportion to that taken in computing, by bank measurement (after clean-up), the cubic content dredged. The results are often vitiated by old workings, clay, etc., encountered by the drill.

No. 7. The method of valuing prospected area may be generally compared to that of valuing a mine from sampling and assaying, except that in the former case the results are not so consistent or accurate. One authority of a very wide experience states that before deducting from the 'prospected' value of the property, the cost of working, interest, depreciation, etc., he reduces the estimated value as obtained from drilling by 40%. As I have already taken considerable of your valuable space, I shall refer to the question of sectionalizing a dredge in another issue.

In the criticism of Mr. Curle's article by C. W. Purington, appearing in your issue of July 28, there are also several points touched upon which, in my opinion, may be themselves criticized. The generalization that "flood plains reduced nearly to base level, in which streams meander are * * * the first requisite for gravels fit for dredging" is disproved by the conditions in the Sacramento valley. Neither the territory on the Yuba where the river bottom, with its accumulation of tailing overburden, forms the gold-bearing material, nor the bench gravels of the American river at Folsom, together forming probably half (in value) of the available fields, come under this head.

The large dredging area adjacent to the American river, near Folsom, seems to be almost ignored by both Mr. Curle and Mr. Purington, though undoubtedly excellent and remunerative work is being done there today.

Referring to Mr. Purington's criticism of question No. 4, I do not understand Mr. Curle's question to refer to every dredge that has *ever been built*, as naturally a dredge's life is only that of the hull, and under early conditions this did not last very long. As far as his remarks refer to the districts of Oroville, Yuba, and Folsom, they seem rather extravagant.

Mr. Purington states that probably not over 100 dredges are working at a profit in the world. As there are probably forty-eight working at a profit in California, and I am credibly informed of at least four others in the United States, the New Zealand practice, where there are over 200 in use, must be a losing business.

APPENDIX. 205

Contrary to the opinion expressed, I contend that shaft sinking is feasible in the large majority of cases, and is much cheaper than drilling. Even where water is encountered the cost in 90% of cases is not increased in indue proportion to results. The question of valuation is one that Mr. Purington covers considerable space in treating, but the sum of his remarks seems to be that "to enquirers like Mr. Curle," at least, no proper system of valuing the ground to show net profits has been discovered. I confess that I cannot see any particular difficulties in computing the purchasable value (which, by the way, is the real test), of a properly examined and prospected dredging area other than those that occur in mine valuation. The increased risk incurred in prospecting dredging ground is offset by the possibility of change in the nature of orebodies with depth, and of necessitating different and more expensive treatment at surface. No competent engineer would value any property, either dredging or lode, without making similar and proper allowances for all expenses, including preliminary and amortization, etc., and therefore I do not see why exceptional stress should be laid on those points when applied to dredging.

D'ARCY WEATHERBE.

Bisbee, October 22, 1906.

Sectional Gold Dredges.

The Editor:

Sir—Continuing the discussion of Mr. Curle's article, I desire to say that the question of sectionalizing a dredge for mule transport over such rough country as the Andes is interesting, and there is no doubt that this may be accomplished, provided timber for the hull and spuds is available on the east side of the range, where a wider road than a bridle-path could be built. As to dividing it up in eight pieces, that is of course out of the question, because the machinery alone for a $3\frac{1}{2}$-cu. ft. bucket dredge would weigh in the neighborhood of 160 tons, exclusive of motive-power machinery. One-eighth of this weight is twenty tons, and two mules, under such circumstances as stated, could not possibly take a load of more than about 600 pounds.

The following particulars of an interesting case of a dredge being sectionalized for a property on the Gold Coast are given me by Mr. C. Cline, formerly connected with the Risdon Iron Works: An English company, known as the Goldfields of Eastern Akim, about the year 1900, gave an order to the Risdon Iron Works, of San Francisco, to build a dredge for transportation from a point near Accra, on the coast of Ashanti, to a point some sixty miles inland, where an alleged dredging tract was situated. The native method, of carrying the load on the head, was to be used, and, as a good porter will only take about 200 lb. in this manner, it will be readily understood that the parts had necessarily to be light. Of course, many pieces could be slung between several natives. It is related that a case is on record of a gigantic native from the Sahara carrying 400 lb. in this manner.

It was intended to construct the hull on the ground, from native timber, so the question of lumber was eliminated. The tumblers were built in twenty-five to thirty pieces each, and as the upper one weighed 4000 lb., this made the heaviest part not over 200 lb. The ladders, trommels, buckets, etc., were naturally all easy to sectionalize. As the dredge was to be run by steam, a boiler was necessary, and the intention was to transport the plates separately, the riveting to be done on the ground. The heaviest single pieces were the engine cylinders, which weighed about 1000 lb. each. The tumbler castings, instead of being pressed onto the shaft by hydraulic press, were to be heated, the expansion being sufficiently great to give a tight fit when cold. At the last moment, when all was ready, orders were received by cable to rivet all sections together and ship as an ordinary dredge. This was done and the machinery was dumped on the beach at Accra. Internal trouble had arisen in the company meanwhile, and there was a stop to further progress. Some of the dissentient members formed a new company and ordered a complete dredge from the Risdon company, which it was proposed to transport by means of traction engines. This machine was also built of $3\frac{1}{2}$-cu. ft. bucket capacity. Accordingly, three years after the first dredge had been shipped, a properly metaled road was built into the property at a cost, it is said, of £30,000. The work was elaborately done, the road surface being actually graveled, though the native carriers unfortunately avoided this and took to the old trail, preferring it because the gravel hurt their bare feet. Three traction engines—each of different design and of uninterchangeable parts—were sent out and proved to be a failure. One, I believe, only reached a point seven miles from the shore. Mr. Cline was finally sent out, and he transported and erected the dredge in six months. All circular parts were cased in wood and rolled along the road. The boiler-drum, to allow for the camber of the road, was built up with wood at the ends and also rolled in. The timber for the hull and framing was cut and framed on the ground, and the stacker-ladder was built of wood. The boiler sections were riveted and the 16-ft. tubes put in on the spot. The ground to be dredged was 14 ft. deep at the place of building and Mr. Cline ran the boat for four weeks after it was erected, but I have been unable to learn of its subsequent fate. As the ground had never been either drilled or, in fact, prospected, I fear that it is not hard to conceive the outcome.

The parts and weights for a $3\frac{1}{2}$-cu. ft. dredge, as usually provided, are given in the following table, with the corresponding number of sections and weights into which each could be divided. The ground, however, to be worked by such a machine far away from proper repair facilities, etc., would require to be extremely rich to allow the feat to be undertaken, and it is doubtful even then whether it would prove practicable.

The following table* will give an idea of the weight of a 3½-cu. ft. Bucyrus dredge and the divisibility of its parts:

Name of Part.	Number of pieces and their weight.
Upper tumbler 6500	Can be cut in 20 pieces, one of which will weigh 1000 lb., the rest will be below 300 lb.
Lower tumbler 4500	Can be cut in 13 pieces, three of which will be about 700 lb., the rest below 300 lb.
Digging ladder 28,000	Two pieces of 600 lb., the rest about 300 lb.
Digging buckets (3½ ft.)....83,000	Bottom about 320 lb., each hood 135 lb., lip 120 lb.
Screen, stacker, and parts...16,000	Eight pieces would weigh about 600 lb. each, all other parts 350 lb. and less; 70% less than 300 lb.
Gearing 30,000	Eight parts would weigh about 700 lb. each, the rest from 350 lb. down; 50% less than 300 lb.
Engine or motors............15,000	Two pieces about 1000 lb.; two pieces about 600 lb.; 50% below 350 lb.
Boilers 8500	All below 350 lb.
Pumps 300	
Winches 42,000	Two pieces 600 lb. All other parts below 350 lb.
Other parts 7600	All below 350 lb.

Bisbee, November 3, 1906.   D'Arcy Weatherbe.

---

* Which I owe to the courtesy of D. P. Cameron, of the Western Engineering & Construction Co., of San Francisco, California.

From Mining and Scientific Press, November 17, 1906.

Cost of Dredging.

The following figures, giving the actual cost of dredging operations on two boats at Oroville, speak for themselves. They are worth more than many estimates:

COST OF DREDGING.

No. 1.

Year.	Cu. yd.	Total cost.	Per cu. yd.
1903	485,016	$38,583.54	$0.07959
1904	466,262	43,430.32	0.09314
1905	352,826	35,699.32	0.10118
1906	465,207	35,401.30	0.07610
Four years	1,769,311	$153,114.48	0.08654

Using 3½-ft. buckets and in easy digging ground.

No. 2.

Year	Cu. yd.	Total cost.	Per cu. yd.
1905	589,082	59,115.10	0.10035
1906	557,084	55,454.00	0.09954
Two years	1,146.166	$114,569.10	0.09996

Using 5-ft. buckets and in hard digging ground.

From MINING AND SCIENTIFIC PRESS, March 2, 1907.

Gold Dredging.

The Editor:

Sir—In your issue of November 3, 1906, Mr. D'Arcy Weatherbe discusses gold dredging, with special reference to Mr. Curle's letter on the same subject in the issue of January 27, 1906. In the closing portion of Mr. Weatherbe's letter, he refers to a letter of mine of July 28, relating to the same subject, and advances some criticisms thereon.

While in no way wishing to decry the valuable contributions of Mr. Weatherbe concerning California dredging, I must admit that I fail to see much force in the last portion of his letter. I have not seen, as yet, from Mr. Weatherbe or others, a specific answer to my question in the issue of July 28, as to the cost of dredging. For the present I leave this point open.

Mr. Weatherbe objects to the "generalization" that flood-plains are the proper field for gold dredging. He answers my statement by an unsupported generalization of his own that "Neither the territory on the Yuba where the river bottom, with its accumulation of tailing overburden, forms the gold-bearing material, nor the bench gravels of the American river at Folsom, together forming probably half (in value) of the available fields, come under this head."

I hope to support, in the following, the view that both the areas referred to are typical examples of "flood-plains reduced nearly to base-level." Before proceeding, I wish to make the comment that notwithstanding the number of highly-trained men now engaged in the examination of dredging conditions, and in the prosecution of the dredging industry, geological factors have not been taken sufficiently into account in selecting the ground. I crave space to insert here once more a description of what I believe to be the essential conditions for a dredgeable territory.

"Taking the country about Oroville, California, as the best example in the United States, let the geological conditions be considered. To the north and east the erosion of a vast extent of mining country, whose rocks are penetrated by gold-bearing veins, has contributed little by little through geological ages to the mass of detritus now occupying the bed of the Feather river. The wearing down of mountains (originally very much higher than at present) through a vast amount of time, has caused the formation of a valley of extraordinary width, but of no great depth. The massing of stream detritus is also responsible for a decrease in the gradient and a slowing down almost to a topographical equilibrium of the formerly swift current of the river. The stream, unable under the conditions to cut its way by virtue of the material which is carried in suspension, has for a long time been depositing its load, filling its wide valley with sand and gravel, together with the less destructible of the metallic particles, and notably gold. Geologically speaking, the rate of deposition of the river's material may be said to be on the increase and the current of the river, still of considerable velocity, to be greatly lessening in swiftness. Later in its geological cycle the Feather river will doubtless assume, on a smaller scale, the present character of the Mississippi, forming ox-bow curves, cut-offs, and as it were, losing its way among constantly shifting sandbars. The accompanying residual gold, as it travels to a greater and greater distance from its original source in the veins of the mountains, becomes more and more finely divided, even to microscopic dimensions, and increases in purity, and the degree of evenness with which the particles are distributed in the gravel becomes a phenomenon of constantly increasing definiteness and importance.

Such a set of conditions as that obtaining on Feather river is the exception rather than the rule in the western part of the United States, and even in California. The above explanation has been entered into in order to present in some measure the reason why, on the California river referred to, the conditions are not only favorable but eminently suited to gold dredging. It is hardly necessary to state that the amount of gold increases in geometrical ratio the lower it lies in a given bed of gravel. Thus, old rather than young valleys are favorable for dredging.

Additional reasons of great weight why geological old valleys should be looked for are that the size of boulders is greatly decreased, gravel becomes by long abrasion uniform in size, the angularity of the fragments disappears, and a bed of pebbles, round and easily handled, is the result. The even distribution of the gold, which, as mentioned above, is an invariable accompaniment of old and wide valleys, is a point in favor of this sort of mining, looked at from the standpoint of a business enterprise. At the same time it is evident that the finely divided state in which such gold is found necessitates the highest skill in recovering it."*

The above description was written before the Yuba and American river areas had become prominent for dredging. These areas are not different in geological history from the Feather river dredging area. In fact, I believe they would be classified by any geologist as part of the same general base-level. The Yuba river, 12 to 16 miles east of Marysville, has a gradient of less than 50 feet per mile. It flows in extensive gravel and silt-filled plain. The fact that 25 feet of tailing from former operations covers the plain in and near the river-bed is an incident of artificial origin, which has accentuated the natural base-level conditions previously imposed. Mr. Weatherbe perhaps means that the gold exists in the bottom of the Yuba river by virtue of the tailing that covers the gravel deposited in the natural way. If this be the case, I admit that a special condition exists, but as it has no bearing on the geological conditions, it does not constitute a refutation of my contention. On the other hand, my understanding has always been that the Yuba gold occurs mostly in the undisturbed gravel beneath the tailing. If this be true, it is as surely the gold of a flood-plain as is that of the Feather river.

The deposition of fine alluvial gold in great river valleys by floods and periodic inundations may be called flood-gold. No one can say in what portions of a given broad valley the local currents will deposit such fine gold. The result must be that the flood-gold occurs in patches distributed irregularly over a wide area. On the Yuba I believe it is a coincidence that one such patch, determined by prospecting, lies in the area crossed by the present river-channel. Similar patches will be found over a greater area than that at present exploited.

As regards depth, wells at Wheatland on the Bear river, a part of the same great base level, went through 150 feet of gravel, clay, hard-pan, sand, and a further 300 feet of greenish sand, all of Tertiary or late Cretaceous age, and it is fair to assume the same depth in the Yuba. Gold may exist throughout this depth in various layers. By any present method it is unrecoverable.

---

*'The Gold-Dredging Fields of Eastern Russia,' by C. W. Purington and J. B. Landfield, *Engineering Magazine*, 1901, p. 398.

Fig. 103. Feather River, with Dredges in Operation.

Messrs. Becker, Lindgren, and Turner, in the Smartsville folio of the Geological Atlas, July, 1895, state, regarding the Yuba valley: "The present system of water-courses is steadily degrading the older formations and depositing the sediment on flood-plains forming bottom-lands of greater or smaller extent or alluvium." The elevation that occurred in Pleistocene time and which affected the American river area thirty vertical feet or more, may have had some influence along the Yuba, and even the very moderate grade of stream was probably less than at present. The river is at present doing but little cutting, except in the way of removing its artificial load. The present dredge-tailing is rapidly removed and re-deposited further down the stream, when the current of the river comes in contact with it. I witnessed an impressive illustration of this cutting away during a season of high water on the Yuba, in October, 1904. As far as the Yuba valley is concerned, I think it would be difficult to find a fairer example of a flood-plained stream reduced nearly to base-level.

The area of bench-gravel now being dredged adjacent to the American river at Folsom offers the conditions of an ancient base-level raised slightly over 30 feet above the present level of the stream that is reducing it. The elevation at Folsom does not exceed 140 feet above the sea, and the gradient of the American river is less than 25 feet per mile. The fact that the auriferous patch of gravel, which is being dredged, lies in a low bench, means in this case that the cycle of base-leveling has been completed, and that the terrane has undergone a slight elevation, and is again being disintegrated. It has not only been reduced to base-level, but has, as it were, been lifted up for exhibition, to show what a base-level flood-plain should look like.

Mr. Weatherbe has shown us that he knows a great deal about dredging operations in California. He will, I am sure, pardon this protest from a geologist who has endeavored to apply scientific principles to the search for dredging territory, and he will recognize that there may be ways of determining what ground presents dredging possibilities, undreamt of in his philosophy.

C. W. PURINGTON.

London, March 11, 1907.

From MINING AND SCIENTIFIC PRESS, April 27, 1907.

INDEX.

	Page
Accident to Bucket-Line of Yuba No. 2 Dredge	103
African Dredging	186
Agricultural Value of Dredging Ground	164
Alaskan Dredging Field	194
Alder Gulch, Montana Dredges.	86
Amalgamating Machine for Clean-up on the Butte and El Oro Dredges	134
American River Bars	15
Average Value of Gravel at Oroville	174
Belt Conveyors	72
Biggs No. 1 Dredge	39
Blue Lead Theory	15
Booming	18
Boston and California Dredge No. 1	38
Bucket-Lines	54, 63
Bucket-Lips	147
Buckets	66, 192
Burial of River-Channels	14
Burmese Dredging	187
Caminetti Act	20
Canadian Dredging	184
Central Gold Dredging Co's. Log Book	32
Centrifugal Pumps	81
China Pump	41, 172
China Shafts	40
Clean-up	128
Clean-up Practice on Yuba No. 4 Dredge	130
Clean-up Practice on Legget No. 3 Dredge	132
Cleaning Mercury	136
Cline, C.	206
Close-Connected Bucket-Lines	102
Conrey Placer Mines	128
Construction of Dredges	46
Core from Churn Drills	28
Cost of Complete Dredge	139
Cost of Drilling Holes	33
Costs of Dredging	159, 197, 209
Curle, J. H.	187
Cutting the Bank	90
Debris Commission	20
Delancy Tract	38

	Page
Deposition of Gold-Bearing Gravels	12, 212
Depth of Dredging	200
Depth of Erosion of Veins	16
Depth of Gravel Deposits	10, 212
Description of Dredges at Oroville	175
Dipper Dredges	84
Dredge Construction	46
Dredging Machines	46
Drill-Hole and Dredge Table	38, 39
Dumping of Buckets	68
Electric Equipment and Transmission	73
El Oro Tract Records of Colors	42
Estimates from Drill-Records	38
European Dredging	187
Expanded Metal in Sluices	110
Factors Affecting Gold-Saving Efficiency of a Dredge	124
Feather River Basalt	14
Financial and Statistical Aspects of Gold Dredging	183, 201
Flotation Level	124
Folsom	26
Folsom Companies	180
Folsom Development Co's, No. 4 Dredge	180
Formation of Terraces, Bars, and Benches	14
Fraser River Dredging	184
Garden Ranch Dipper Dredge	84
Gauntrees	52
Gearing for Bucket-Line Driving	78
Geological Sketch of California	9, 210
Glacial Period	12
Gold-Saving Apparatus on Folsom No. 4 and 5 Dredges	123
Gold-Saving Efficiency of a Dredge	124
Gravel in Tertiary Rivers	10
Grizzlies	110
Ground Sluicing	18
Hammon City Dam	21
Hammon and Treat	40
Hammon, W. P.	25
Head-Line Anchorage	90
Historical Sketch	10

INDEX.

	Page
Holmes, George L.	73, 193
Holmes' System of Launders and Tables	118
Horticulture and Dredging	164
Hulls of Dredges	46
Hydraulic Elevator	20
Hydraulic Mining	20
Ideal Dredging Conditions	184
Insulated Cables	108
Jets of Water for Emptying Buckets	69
Keystone Drill	27
Known Areas of Dredging Ground	184
Ladder on Folsom Dev. Co's No. 5 Boat	96
Ladder Rollers	152
Ladders	54
Lava Flows	14
Life of Conveyor Belts	154
Life of Pins on Butte Dredge	150
Lips of Buckets	145
Loss in Digging Time	141
Loss of Gold in Dredging	127
Lost Time on Butte, El Oro, and Lava Beds No. 3 Dredge	142
Lost Time on Exploration No. 1 Dredge	143
Lost Time on Yuba No. 1 and No. 2 Dredges	144
Marine Idea	15
Marion Steam Shovel Co. Buckets	147
Marysville	25
Matto Grosso Gravels	186
Melting Room Practice	136
Method of Extracting Pins	151
Mooring Boats	83
Motors on the Gaggette Dredge	176
Motors on the Boston No. 4 Dredge	177
Nome Dredging Fields	195
Number of Dredges in the World	197, 204
Number of Holes Necessary in Testing Ground	33
Ogilvie's Dredge	193
Open Bucket-Lines	102

	Page
Oroville Companies and Dredges	170
Oroville Dredges Described	175
Oroville Gravel Value and Costs of Working	201
Paddock System	40
Pan	16
Pins	66, 148
Post-Glacial Period	14
Prehistoric Man	14
Preventing Salting	30
Production of Gold	16, 196
Prospecting Dredging Ground	28, 198
Prospecting with Churn Drills	28
Purington, C. W.	199, 214
Ratio of Recovery Between Dredging and Drilling	36
Reclamation of Worked Ground	165
Repair Shops for Yuba and Folsom Dredges	158
Rhodesian Dredging	186
Ridgway Conveyor-Belts	155
Robinson, A. Wells	193
Rocker	17
Russian Dredging	195
Rusty Gold	127
Sacramento Valley	24
Sacramento Valley During Glacial Epoch	12
Salting Drillings	31
Saskatchewan River Dredging	184
Save-All Sluices	112
Screens and Riffles on Leggett No. 3 Dredge	116
Screens; Duty, Cost, and Repairs	109
Secondary Concentration	14
Sectional Dredging Machinery	190, 192, 206
Shaft-Sinking for Testing	40, 205
Shafting Sizes	63, 192
Single-Lift Dredges	86
Source of Channel Gold	16
South American Dredging Fields	185, 196
Special Labor	162
Spuds	93, 155
Stackers	71
Steps in Placer Mining	16
Tables	110
Tables and Sluices on Biggs No. 2 Dredge	115

INDEX. 217

	Page
Tables and Sluices on the Pennsylvania Dredge	114
Tables and Sluices on Yuba No. 4	114
Tertiary River Channels	10
Theories of Source of Gold	15
Tom	18
Traveling Cranes	155
Tumblers	57, 192
Value of Ground in Different Districts	137
Valuing Dredging Ground	27
Variable Flotation Level	124
Volcanic Activities	10
Wear on Buckets	145
Weatherbe, D'Arcy	205, 208
Weight of Sectional Dredge	208
Yuba Bottom	178
Yuba Companies	177
Yuba Dam	22
Yuba River Bars	14

THIRTEEN FOOT PLACER DREDGE
AT FOLSOM, CALIFORNIA.

The machinery for three-fourths of all the modern placer dredges now in use in the United States and Alaska, in addition to all the steam shovels used in digging the Panama Canal, has been furnished by this Company. - - - - -

The Bucyrus Company
South Milwaukee, Wisconsin.

GOLD DREDGES

WORKMANSHIP, WEIGHT AND MATERIAL.
UNEQUALED.

We build only Gold Dredges.   We make prompt deliveries.
We design and build each and every dredge to fit the particular conditions under which it is to operate.
Our location commands the Export work for South American countries.

New York Engineering Company

United States Express Building

2 Rector St., New York, N. Y.

In writing for catalogue give description of your proposition.

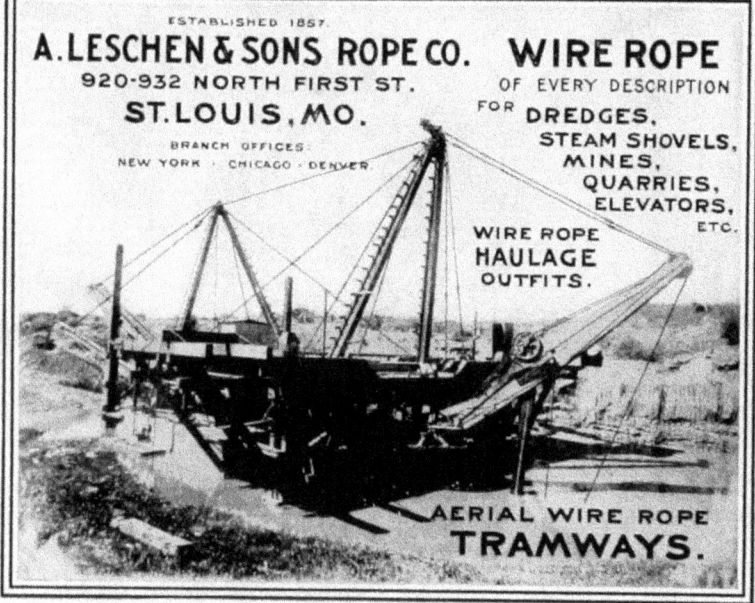

Keystone "Churn" Drills

--FOR--

PLACER GOLD TESTING

COPPER AND ZINC PROSPECTING

PERCUSSION COAL CORING TESTS

WATER WELL DRILLING

OIL AND GAS WELL BORING

HEAVY BLAST HOLE DRILLING

Our five catalogs are scientific treatises on these subjects. Catalog No. 2 deals with mineral tests. All free.

KEYSTONE PLACER DRILL CO.
BEAVER FALLS, PA.
New York Office: 170 Broadway.

San Francisco:
Harron, Rickard & McCone.

Seattle:
Caldwell Bros. Co.

The Metallurgy of the Common Metals

Gold, Silver, Iron, Copper, Lead, and Zinc.

BY

Leonard S. Austin.

Price $4, postage prepaid.   1st edition 1907.

Published and for Sale by

Mining and Scientific Press

667 Howard St., San Francisco.

ESTABLISHED MAY 24, 1860.

T. A. RICKARD, Editor.   EDGAR RICKARD, Bus. Mgr.

The MINING AND SCIENTIFIC PRESS offers every Saturday THE LATEST ACCURATE MINING NEWS, written by paid correspondents in every mining centre of the world.

VALUABLE AND TIMELY DISCUSSIONS on problems daily facing the man in charge of the mine, mill, and smelter.

ORIGINAL ARTICLES on the most recent practice by leading members of the mining profession.

$3 PER ANNUM
ADD $1 FOR POSTAGE TO CANADA
ADD $2 FOR FOREIGN POSTAGE

667 HOWARD STREET, SAN FRANCISCO

NEW YORK	CHICAGO	DENVER
42 Broadway	934 Monadnock Block	420 McPhee Building

LONDON: LEATHWAIT & SIMMONS
5 Birchin Lane.

www.ingramcontent.com/pod-product-compliance
Lightning Source LLC
Chambersburg PA
CBHW071714160426
43195CB00012B/1671